John Marie Keating

Maternity, Infancy, Childhood

Hygiene of pregnancy; nursing and weaning of infants

John Marie Keating

Maternity, Infancy, Childhood
Hygiene of pregnancy; nursing and weaning of infants

ISBN/EAN: 9783337373399

Printed in Europe, USA, Canada, Australia, Japan

Cover: Foto ©berggeist007 / pixelio.de

More available books at **www.hansebooks.com**

PRACTICAL LESSONS IN NURSING.

MATERNITY, INFANCY, CHILDHOOD.

HYGIENE OF PREGNANCY; NURSING AND WEANING OF INFANTS; THE CARE OF CHILDREN IN HEALTH AND DISEASE.

ADAPTED ESPECIALLY TO THE USE OF MOTHERS OR THOSE INTRUSTED WITH THE BRINGING UP OF INFANTS AND CHILDREN, AND TRAINING SCHOOLS FOR NURSES, AS AN AID TO THE TEACHING OF THE NURSING OF WOMEN AND CHILDREN.

BY

JOHN M. KEATING, M.D.,

VISITING OBSTETRICIAN AND LECTURER ON THE DISEASES OF WOMEN AND CHILDREN, PHILADELPHIA HOSPITAL (BLOCKLEY); FELLOW OF THE COLLEGE OF PHYSICIANS OF PHILADELPHIA, ETC.

SECOND EDITION.

PHILADELPHIA:
J. B. LIPPINCOTT COMPANY.
LONDON: 10 HENRIETTA STREET, COVENT GARDEN.
1890.

Copyright, 1887, by J. B. LIPPINCOTT COMPANY.

TO

J. M. DA COSTA, M.D., LL.D.,

PROFESSOR OF THE PRINCIPLES AND PRACTICE OF MEDICINE, JEFFERSON
MEDICAL COLLEGE,

IN RECOGNITION OF MANY ACTS OF
PERSONAL KINDNESS.

PREFACE.

THIS little work is intended for mothers, and for those who have undertaken the care of infants and children in health and sickness. The enormous mortality of children under three years of age, the greater part of which is preventable, has attracted the notice not only of physicians, who have long ago insisted that the methods adopted in general for the weaning of children were unscientific, but of the public also, who are beginning to realize this fact, and now willingly accept the advice of those who are giving especial attention to the subject.

Much of the material appearing in this book has been taken from the manuscripts of my lectures to students of medicine and also pupils of training schools for nurses.

I sincerely hope that the practical nature of many of the suggestions offered will find favor with the busy practitioner of medicine, and be looked upon as a supplement to the text-books on children's diseases that cannot give space to details of nursing.

To Dr. Charles S. Turnbull I am indebted for the article on the care of the ear and eye, one of the most important chapters in the book.

Dr. Alexander W. MacCoy kindly replied to some questions, and the matter is so practical and clear that I have placed it in a chapter as sent me.

1504 WALNUT STREET, PHILADELPHIA.

CONTENTS.

PART I.
MATERNITY.

CHAPTER I.
PREGNANCY.

How to Calculate the Probable Date of Confinement—The Presumptive Evidences of Pregnancy—Morning Sickness; the Cause, and how to Remedy it—How to Avoid and Remedy the Constipation of Pregnancy—Exercise, and how it should be taken—Diet during Pregnancy—The Care which the Breasts should receive—The Baby-Basket—A List of Articles necessary in Confinement—The Binder—Miscarriage; what is Meant by the Term—The most Frequent Causes—The Greatest Care and Caution should be Exercised after Miscarriage 13

PART II.
INFANCY.

CHAPTER II.
CARE OF NEW-BORN INFANT.

How to cleanse a New-born Babe—The Cord, and the Care which it should receive—The Binder and its Object—Importance of noting the Child's Secretions—A Babe should be given Water occasionally—All about the Mother's Milk first making its Appearance—The Diet of the Nursing Mother—How to check an excessive flow of Milk, and to increase an Insufficient Supply—The Child's Diet—How often a Child should receive Nourishment 38

CHAPTER III.
BOTTLE-FEEDING.

The Child's Bottle and how to prepare it—Great Care in preparing the Bottle must be Exercised—An Efficient Nurse indispensable to the Welfare of the Child—A Study of Milk—Condensed Milk and fresh Evaporated Milk 50

CHAPTER IV.

PREPARED BOTTLE.

Cow's Milk, its Advantages and Disadvantages—A Child's Digestion—Different Preparations of Bottle-Food—Why an Infant should have very Little Starchy Food in its Diet—How to avoid either Constipation or Diarrhœa—A Baby's Bottle Diet may be varied—Peptonized Milk—The Care which a Mother should exercise in the Selection of a Child's Diet 66

CHAPTER V.

WEANING.

Weaning—When to wean a Child—Where to wean a Child; and how to wean it 83

CHAPTER VI.

FRESH AIR, VENTILATION, OUT-DOOR EXERCISE.

The Great Importance of Ventilation—The Selection and Care of the Nursery—How it should be Heated—The Danger of allowing Children to be left to the Care of Young and Inexperienced Girls—How Children frequently contract Diseases 87

CHAPTER VII.

BATHING.

The Importance of Bathing—A Child's Time for Bathing must be regulated by the Child's Condition, etc.—The Cruelty of forcing Young Children to have a Plunge in the Cold Sea—A Child should love its Bath, and how it can be taught to love it 94

CHAPTER VIII.

TEETHING.

The Milk-Teeth, and when they are usually Cut—Why a Slobbering Babe suffers less from its Teeth than another—Usual Symptoms—Why Mothers dread a Child's Second Summer—Rickets and what is meant by the Term—Rickets a Frequent Cause of Spinal Affection—How a Pregnant Woman may predispose her Child to Rickets—Fat Children are not always healthy ones—How to treat a Rickety Child—A Teething Child suffers frequently from other Causes than its Teeth—A Child should have Sufficient Sleep—Why so many Little Ones are wakeful—How Sleeplessness can be remedied—The Diet of a Teething Child 100

CHAPTER IX.
DIET AFTER EARLY DENTITION.

The Necessity of a Careful Selection of Diet—The Advantages and the Disadvantages of taking a Child to the Table—The Necessity of discriminating between Children as to their Diet—Why what is Meat to one Child is Poison to Another—Exercise is Essential 112

CHAPTER X.
ON THE BOWELS.

What constitutes a Normal Condition of the Bowels in a Child—More Grease than Powder should be used for Young Children—What causes Disturbances of the Bowels—Different Causes of Constipation—How Constipation can be overcome—Of what a Child's Diet should consist, and how it can be regulated when it is suffering from Bowel-Complaints—A few Useful Remedies—Exercise a Specific for many Forms of Constipation 116

PART III.
CHILDHOOD.

CHAPTER XI.
ACUTE AND CHRONIC NASAL CATARRH.

Affections of the Mucous Membrane of the Nose, Acute and Chronic, in Infants and Children—Their Prevention and Treatment 127

CHAPTER XII.
DISEASES OF THE EAR AND EYE.

Diseases of the Ear in Infancy and Childhood—The Care of the Ear in Childhood—Diseases of the Eye in the Newborn—Treatment of Simple Ophthalmia—Contagious Ophthalmia; how to prevent it; its Treatment and Nursing . 135

CHAPTER XIII.
DISEASES OF THE THROAT AND AIR-PASSAGES.

Croup and Diphtheria—Simple Spasmodic Croup, what its Symptoms are and how to treat it—Membranous Croup, its Symptoms and Treatment—The Difference between Membranous Croup and Diphtheria—Diphtheria as a Cause of Membranous Croup—The Nursing, and the Use of Household Remedies in their Treatment 151

CHAPTER XIV.

DIARRHŒA.

The Causes of Diarrhœa—Over-feeding; Tainted Milk; Decomposition of Food; Undigested Starches; Teething; Hot Weather—Inflammatory Diarrhœa—How it can be avoided—Change of Diet necessary, also Absolute Quiet, Pure Drinking-Water, and Fresh Air—The Character of the Diet—Importance of Peptonized Milk and Nutritious Injections 164

CHAPTER XV.

WHOOPING-COUGH.

Its Character—Its Complications—The Nursing of it—Diet—Treatment 174

CHAPTER XVI.

SCARLET FEVER.

Its Cause—The Reasons why it does not appear to be as Contagious as other Similar Affections—The First Symptoms—Its Nursing—Complications—Sequelæ 180

CHAPTER XVII.

MEASLES.

How it is contracted—How the Contagion is carried—Why it is the most contagious of the Eruptive Diseases—The Peculiarity of the Eruption—The Dangers of Pulmonary Troubles as Complications or following the Disease—The Nursing 186

CHAPTER XVIII.

SECOND DENTITION.

Forcing in Education to be Condemned—What is Meant by Second Dentition—Complications Due to Hereditary or Acquired Conditions—Rickets—How it is to be Avoided—Diet of Children at this Age—Clothing 191

CHAPTER XIX.

PUBERTY.

Puberty—The Four Second Molars—The tendency to Disturbance of Digestion at this Time—The Importance of Proper Food and Clothing, also Freedom from Excitement and the use of Iron—First Menstruation—Menstrual Irregularities—Hysteria—Dysmenorrhœa—The Abuse of Anodynes—The Mother the proper Confidant of her Daughter. 200

MATERNITY; INFANCY; CHILDHOOD.

PART I.
MATERNITY.

CHAPTER I.

PREGNANCY.

How to Calculate the Probable Date of Confinement—The Presumptive Evidences of Pregnancy—Morning Sickness; the Cause, and how to Remedy it—How to Avoid and Remedy the Constipation of Pregnancy—Exercise, and how it should be taken—Diet during Pregnancy—The Care which the Breasts should receive—The Baby-Basket—A List of Articles necessary in Confinement—The Binder—Miscarriage; what is Meant by the Term—The most Frequent Causes—The Greatest Care and Caution should be Exercised after Miscarriage.

CONCEPTION is more liable to take place either immediately before or immediately after the period, and, on that account, it is usual when calculating the date at which to expect labor to count from the day of disappearance of the last period. The easiest way to make a calculation is to count back three months from the date of the last period and add seven days; thus we might say that the date was the 10th of June: counting back three months brings us to the 10th of March, and adding the seven days will bring us to the 17th.

Very many medical authorities, distinguished in this line, have stated their belief that women never pass more than two or three days at the most beyond the forty weeks conceded to pregnancy,—that is, two hundred and eighty days or ten lunar months, or nine calendar months and a week. About two hundred and eighty days will represent the average duration of pregnancy, counting from the last day of the last period. Now, it must be borne in mind that there are many disturbing elements which might cause the young married woman to miss a time. During the first month of pregnancy there is no sign by which the condition may be positively known. The missing of a period, especially in a person who has been regular for some time, may lead one to suspect it; but there are many attendant causes in married life, the little annoyances of household duties, embarrassments, and the enforced gayety which naturally surrounds the bride, and these should all be taken into consideration in the discussion as to whether or not she is pregnant. But then, again, there are some rare cases who have menstruated throughout their pregnancy, and also cases where menstruation was never established and pregnancy occurred. Nevertheless, the non-appearance of the period, with other signs, may be taken as presumptive evidence.

The exact date at which to expect confinement is an important one to determine for many reasons; the engagement of a nurse is dependent upon it, the choice of a nurse being a matter of importance. Great difficulty attends the positive diagnosis of pregnancy until the fourth month, although there are many presumptive evidences of its existence. Among these may be noted the changes in the breasts, their increasing size and fulness; they become larger and harder, the areola (the dark line which surrounds the nipple) becomes increased in size and darker in color, especially in brunettes; later the nipple itself becomes larger in size and darker in color, and here and there in the dark ring surrounding it will appear small tumors or indurations, that stand out very prominently : these are glands. In the mean time the breasts themselves become very sensitive, there is a feeling of fulness, the clothing fits tightly, and there is a sense of congestion; these changes may take place without the existence of pregnancy,—such cases, however, being rare,—but these signs may be ranked as presumptive evidence, particularly if they follow the absence of a period and make their appearance about the sixth week. The time when undoubted signs of pregnancy appear is towards the end of the fourth month, and even then it would require the greatest caution on the part of the physician as to a positive declaration of its existence,—until the sound of the infant's heart is heard distinctly and apart from the pulse of the mother; soon after this follow the movements of the child, which is called quickening.

At this period the timid and anxious wife develops feelings of a varied character; doubts will arise as to what the future will bring forth : will she be a fruitful woman in reality, or will she be destined to go through life without the sacred name of mother? Strange emotions will be aroused,—that peculiar, mysterious dread, which is an admixture of fear and longing, possibly only appreciated by those who have gone through a like experience. With it all, she secretly welcomes, with all the fulness of her heart, the child who is slowly developing within her, receiving nourishment from her blood and dependent upon her for existence; but she dreads the ordeal through which she must pass before pressing that infant to her bosom. Then, too, the uncertainty of her life and that of her offspring give her many hours of anxiety. Impressions from very slight causes are made more deeply than at any other time; casual remarks, revolting sights, histories of horror, all leave an indelible stamp upon her sensitive nervous system, and in silence they come before her, steal upon her unawares, and require the strongest effort of the will to be overcome. It is right for us to discuss this matter freely with her, to give her a few words of sympathy and caution,—words of sympathy, to sustain in those lonely hours when doubt and fear are crowded upon the young wife who has no one in whom she can confide. Let her take courage; she is not alone in her trouble; hundreds of her sex, who one year ago were haunted by the same visions, troubled by the same uncertainties and doubts, are to-day as happy mothers as their fondest wishes could have enabled them to be.

Pregnancy is not a disease, it is a normal function of woman, and this should be impressed upon the young wife that she may undertake its duties and responsibilities with a thorough knowledge of its requirements; that she may submit wisely to the laws of hygiene, and make the proper preparation for the event which she naturally dreads. It is the natural function of woman to bear children, and nature endeavors to make all her functions normal; and diseases or disorders of various kinds are usually brought about by something which is controllable or avoidable. We would not have it supposed for one moment that we propose to lay down certain cast-iron rules to be followed in every case. There being all kinds and conditions of women, differences existing as to face, temperament, physical health, social position, as well as financial status, a rule which might govern one in her diet, occupation, and general hygiene, might not govern another. The object of this work is simply to state as clearly as possible, in language devoid of technical expressions, all that is of interest to a mother, and it must be, of course, limited to those suggestions which will be of service in ordinary cases. All matters that differ from what is here written should be referred to the family physician for his advice. My own experience leads me to know that there are hundreds of things which a patient forgets to ask her doctor, or thinks too trivial to consult him about, which she should know, or she may hesitate because she thinks it may show unparalleled ignorance on her part, and she dislikes to confess it. Having read all, then, that is contained in this little work on the subject which interests her for the moment, should she desire further information her family physician is the one to supply it.

It would be difficult to place in order all the important means to secure health during pregnancy, but we may sum them up in the statement that *moderation in all things is a fundamental law of hygiene.*

There are many ways by which the physician can establish pretty conclusively the existence of pregnancy after the second or third month. After it has been fully determined that pregnancy exists, great care should be taken to give the child every opportunity to reach full term. A moment's thought will convince any one that the close relationship existing between mother and child, which continues up to the time of its birth, not only influences its growth and development, but also, through its close relationship, impresses upon it certain mental characteristics.

In the earlier months of pregnancy the digestion becomes difficult; in many cases the appetite is either lost altogether or perverted at times; or there is a capriciousness which will give rise to an inordinate desire for food of a kind which may not be considered nutritious. There seems at times to be a demand for certain elements, owing possibly to the withdrawal of them in behalf of the child, which in the process of growth needs these various articles to form certain tissues.

When on the subject of Rickets, I shall dwell at length upon this matter, because it is a disease of the child which proceeds from a deficient supply of lime in its constitution; the child frequently needing the lime to form its bones, and if it does not get it from the food which is supplied to the mother, will withdraw it from her own tissues. It is scarcely necessary to call attention to the fact that the lower animals are in constant search of food that will supply these cravings of nature. The character of the egg laid by the hen will depend upon the kind of food she gets; she will seek lime for the shell, and if she does not succeed in getting it the shell will be soft or thin. There are many curious records of the craving of women for articles of food, and if possible, when such cravings exist, they should be gratified within reasonable limits.

Women have been known to eat earth for the lime. This craving for lime exists more especially toward the latter part of pregnancy, at which time food should be changed accordingly.

In the ordinary mixed diet to which we are accustomed, a large amount of bone-forming material is always present; elsewhere in this work I shall explain the whole subject of digestion, in order that the mother may have a clear understanding of the subject with reference to the feeding of the child. Indigestion should be avoided. This is, as we all know, easily caused by late suppers, eating when one is fatigued, a heavy breakfast in which hot cakes are one of the principal articles, the excessive use of tea or coffee, and similar indiscretions. The extent to which one may indulge depends upon an individual law for each individual case; and we all know, without any deep study, exactly how much and what we can eat, but I fear we are frequently but too willing to run a little risk.

In the earlier stages of pregnancy there is, as we have before stated, frequently a disturbance in the digestive functions, which gives rise to a great deal of annoyance, and at times may become quite a serious consideration; particularly if it prevents the taking of nourishing food, or produces strain from excessive vomiting.

In many instances it does not need the actual presence of food in the stomach to produce it; it comes on at the earliest waking, and is often compared to the sensation of nausea which we have all experienced from want of food. At other times it seems to arise from the presence of food in the stomach, and the breakfast is no sooner taken than violent nausea and vomiting follow; the matter expelled being extremely bitter, due possibly to a certain amount of bile collected in the stomach.

The first food entering the stomach will likely be rejected; but after being relieved, a meal will then be retained with perfect ease and comfort. Though called morning sickness, this nausea differs with different individuals; some complaining of it in the evening and others in the earlier hours of the day.

There is frequently in connection with this disturbance the regurgitation of intensely-acid, burning, watery material, which is very annoying, and is usually accompanied by heartburn. This can be held in abeyance by some alkali, particularly one which has a laxative effect; from one-half to one teaspoonful of light magnesia, taken in water at bedtime, the frequent use of bicarbonate of soda, or the preparation known as soda-mint, either in solution or the compressed pills, will give relief. When not contraindicated by persistent constipation, lime-water and milk, in the proportions of one tablespoonful of lime-water to an ordinary glass of milk, may be taken freely with very good results. Sometimes certain odors will induce nausea: they may be disagreeable ones, as those arising from cooking, and again the most fragrant flowers may produce the same effect. So apt is this to occur that the odors of various plants, which are at other times agreeable, will during the earlier months of pregnancy be decidedly nauseating, and frequently be the first sign of the existence of this condition. Not only does the sense of smell show extreme delicacy as the result of this reflex stimulation, but all the other organs of special sense may be equally affected; for instance, the taste, eyesight, or hearing may be subject to the same disturbances.

The nausea in ordinary cases usually passes off by the third or fourth month; sometimes it lasts longer, frequently continuing until labor sets in. It is one of the most distressing symptoms of pregnancy, and this, with the heartburn which usually attends it, often resists all the means to which one would resort under ordinary circumstances. Should this be the case, consult your family physician as soon as possible.

For morning sickness we may mention the value of some light nourishment in small quantities taken either before or immediately after first rising; such as a cup of chocolate, coffee, cocoa, a cup of good broth well seasoned, a glass of champagne, or brandy and cracked ice; a seidlitz powder or half of one even, or a claret glass of Congress water, taken before rising, will frequently control the nausea. Should there be excessive heartburn, from fifteen to twenty drops of aromatic spirits of ammonia in a wineglass of water, or a mixture of equal parts of lime-water and milk, will have a very beneficial effect.

Frequently five drops of the wine of ipecac in a wineglass of water, given at intervals of from ten to fifteen minutes, will check the nausea.

As this nausea is not dependent upon any disturbance in the stomach itself, but is reflex, it can frequently be avoided by attending to the precautionary measures of which we will make mention.

The patient must endeavor, notwithstanding the unsettled state of her stomach, to take sufficient nourishment during the day, even at this early period. Very often a brisk walk, a little wine or cordial, or a glass of malt extract, which is both nourishing and stimulating, or some preparation of pepsin, as lacto-peptine, for instance, if taken before a meal will cause it to be retained. Should there be craving after an unusual article of diet at any time, the matter should be referred to the physician, and possibly the substance which nature seems to demand may be supplied by medicinal means, or the diet be so arranged as to include those articles which would at first seem questionable, as pickles, etc.

DIET AND HYGIENE.

It is well that the patient should have regular habits from the very commencement of this condition, if possible. I would not have it supposed, however, that she must be restricted to those subjects mentioned in the limits of this article. A few hints may be of advantage as suggesting the proper course to pursue.

Upon awaking in the morning, a small cup of black coffee, cocoa, or chocolate should be taken before rising; she should then take a bath or sponging, to invigorate the system and restore circulation; if possible, she should then take a short walk before breakfast.

One is frequently asked whether a bath is not contraindicated at this time. By a bath we do not mean a plunge into ice-cold water. Cleanliness is never contraindicated under any circumstances. It is as important to have a full action of the skin as of the kidneys or bowels. A bath of a temperature of 75°, taken in a warm room, followed by a gentle surface rubbing or a gentle massage to the limbs, would be of great service and should be insisted upon; but if a bath cannot be taken, a rapid sponging of the body should not be omitted. There are some persons who find it suits them better to take a bath at night; this is not a mere matter of fancy, and I think can be safely left to the patient's judgment. As for hot baths, except those used occasionally simply for the purpose of cleanliness, unless followed by a cold douche they are beyond a doubt debilitating, and the sudden action of cold water might be attended by serious results. We simply warn against extremes.

At breakfast the patient should take either a broiled mutton-chop, one that is thick and juicy, or a piece of tender steak, bread and butter, with one of the cereals, either oatmeal or cracked wheat, in which the whole grain is crushed, or yellow corn; these can be varied according to fancy. Eggs and oysters should be used freely; if possible a pint of milk, or cream, should be taken at each meal; this will supply bone, muscle, and nerve, and at the same time nourish the child,—not at the expense of the mother's own strength and depriving her of the materials which are essential to her own well-being.

The dinner, or heavy meal, should be taken in the middle of the day; in the evening the patient should take only a light meal, so as to relieve the stomach of its digestive burden before bedtime. It is of great importance that there should be the utmost regularity in taking meals, and equally so that there be a diversity in the cooking. Nature requires a change; condiments stimulate the digestive functions, which are apt to become sluggish owing to the fact that the stomach is frequently overloaded. The action of the liver being interfered with, the bowels become sluggish, attacks of biliousness supervene, and the patient loses her appetite and becomes emaciated.

There are several things which we should take into consideration in discussing the dietary in the pregnant state: one is the free use of fruit and vegetables, and the other, the value of water to keep in solution those materials which are intended to be thrown off,—the ashes of the food, as it were, and the waste resulting from the destruction of tissue. For an average individual, taking a fair amount of exercise, five pints of fluid should be taken in twenty-four hours, to make up for the loss of that which passes away by the kidneys, bowels, skin, and breath; this is in addition to about fifteen ounces of fluid contained in the solid food. It is for this reason that milk should form a large portion of the diet of a pregnant woman, and that she should eat freely of fruit, which, as is well known, contains a large amount of water. If such a diet were more generally adopted, the use of purgatives would not be required.

Should the dinner-hour be late, it is advisable to relieve the fatigue incident to household duties, or the usual shopping and visiting, by taking a mid-day meal consisting principally of milk or some light farinaceous food, a milk-punch with an egg beaten up in it, a few biscuits with a glass of Hoff's malt,—something light and easily digested,—after having taken a nap, in order to avoid the great fatigue's interfering with digestion. This should be done invariably.

CONSTIPATION.

One of the most important matters for our consideration is that referring to the regulation of the bowels. Habit, of course, is the best laxative upon which one can depend; no duties, however pressing, should at any time interfere. It is usually the custom for persons to have a movement of the bowels after the first meal in the morning; the presence of food or warm drinks in the stomach acts as a reflex stimulant.

With many, the hour that is chosen is just before retiring; the presence of a full rectum crowding the contents of the abdomen, and the gases which are formed during the day, making the patient restless and uncomfortable. In the earlier stages of pregnancy, constipation is very apt to be obstinate and annoying, owing to the fact that the womb becoming heavier sinks at that time upon the rectum and causes accumulation to take place.

After the fourth month constipation is not so prevalent; indeed, not until the eighth month, or towards full term, does it again become annoying in many cases. Constipation may be produced by two conditions, one the mechanical pressure, the other by what ordinarily is known as biliousness, which in reality is deficient action of the liver and impairment of its function, by which the bile is either not secreted or else it fails to be passed into the bowel. In individuals who take little exercise, no fresh air, and eat heavily, even should that eating consist only of extremely nutritious food, very readily digested, the liver has more than it can possibly manage, the bile is not properly made, and the symptoms of biliousness occur, with constipation. The tongue becomes coated, the breath heavy, the urine is high-colored, the patient complains of loss of appetite, sleeplessness, and irritability.

The bile has various uses independent of its action in digestion; it is a purgative of itself, and when flowing freely causes regularity in the bowel; it is an antifermentative, and prevents flatulence. Constipation which is the result of so-called biliousness demands a variety of measures by way of treatment: less solid food, more fresh air, the free use of water,—pure water.

It is a great mistake, and a very serious matter, for a person who is pregnant to use strong purgatives, unless it be by the advice of a physician. The bowels should be moved by habituating herself to a certain hour, by a moderate amount of exercise, by a glass of water or a cup of chocolate or coffee taken the first thing in the morning, by eating fruit, raw or cooked, or by taking a glass of water at night before retiring. The simplest means should always be used first. Beef-tea or chicken broth, well salted, when taken before retiring, is quite often laxative.

If these fail, the compound liquorice powder, a teaspoon or dessertspoonful, can be taken in a little water or milk. A wineglass of Hunyadi water taken in the morning, two or three times a week, a seidlitz powder, or the patient may prefer a glass of Congress water; when cold and sparkling it is not unpleasant to take, arrests nausea, and acts gently upon the bowel. The elixir of cascara sagrada, a teaspoonful taken at night, should other things fail or become monotonous, or a teaspoonful of maltine, with cascara, taken in the same way, is often efficacious. I consider it unwise to suggest any form of pills for the patient in this state to take without consulting the physician.

Constipation sometimes causes swelling of the left leg or enlargement of the veins, which disappears when the bowels are regulated. The family physician should always have his attention called to any swelling of the body, be it the hands, feet, or face, as it may be a very important matter.

Very frequently it will be necessary for the patient to take an enema. After the fourth month, this should be done at least once a week; before that time it is well to consult the physician, for should there be any evidence of threatened miscarriage it might be productive of harm. It is not at all essential that a large quantity of water be injected into the bowel in these cases in order to get a movement: one or two teacupfuls of warm Castile soapsuds, or simply warm water with a pinch of salt, will be all that is required. The gluten suppository made by the Health Food Company acts admirably in many cases, the object being simply to avoid all straining. Under no circumstances should a strong effort at bearing down be made. The Davidson syringe is probably the most convenient to use, though a fountain syringe often serves better.

Mr. Snowden, an instrument-maker of Philadelphia, has made for me what we call a traveller's syringe. It is made on the principle of an ordinary rubber flower-sprinkler. It can be fully charged with whatever materials are to be used, and put aside until the most convenient time to use it. The patient can readily use it herself, which is a great advantage.

The best time for an enema is probably upon retiring. The weight in the abdomen, and the displacement of its contents, owing to the increasing size of the child, causes pressure on the bladder, and frequently gives rise to distressing symptoms, with a desire to pass water; of course these cannot be relieved entirely, they must be recognized as physiological. Standing much upon the feet or walking is apt to increase this, therefore the recumbent position, for a few moments at a time, is the most desirable, in addition to relieving the bladder frequently. Indeed, during the earlier months of pregnancy this should be attended to with great regularity, for often thoughtlessness, or possibly embarrassment, will permit an over-distended bladder to press the womb out of its position, and possibly give rise to a miscarriage.

The pressure of the womb and its contents upon the blood-vessels in the abdomen produces at times not only disturbances in the action of the liver, but also interferes with the circulation through the kidneys, and if allowed to continue without proper medical treatment might produce serious disturbance; for this purpose it is usual for an examination of the urine to be made frequently during the course of pregnancy. There is also a variety of distressing symptoms produced by pressure on the larger nerves, especially the sciatic nerve, as pregnancy advances; this is particularly noticeable a few weeks before the full term, although it may occur at any period. Those who swim know the disagreeable sensation caused by sudden cramp in one of the limbs. This same cramp, to a greater or less degree, may come on while the patient is sitting or walking, and be productive of great pain; it is usually felt down the back of the thigh, in the foot, or down the leg, especially on the left side; this is greatly relieved by rubbing with chloroform liniment, but the most rapid means is for the patient to assume what is known as the knee chest position,—getting on her knees in bed, with her head on the pillow, thus throwing the weight of the child forward. The greatest relief follows this position, if taken upon retiring at night, in many cases also where there has been distressing flatulence.

A FEW WORDS IN REGARD TO EXERCISE.

It is scarcely necessary for me to dwell long upon the importance of exercise in the open air for giving an appetite, in fact, for the maintenance of health; but just how much exercise to take is often a difficult matter for one to determine. The muscles should be made strong and healthy, the body erect, the circulation free in all parts of the body, and at the same time the exercise should cease before the sense of fatigue is so great as to cause prostration. Then, again, the exercise should be of a pleasant nature; if possible, the patient should be entertained; she should have a purpose in view. The habit of strolling listlessly from place to place simply for the sake of fresh air falls far short of giving the benefit which a brisk walk or a healthy, pleasant occupation would bring about. We all recognize this fact, and we all know the healthy individual is the one who is occupied, whether the occupation be household duties, which are in themselves by no means devoid of active exertion, or whether it be the obligatory hard work of the laboring woman whose robust health will enable her to bear her child with ease compared with the labor of the delicate, weak-backed girl who is brought up in the lap of luxury with every wish gratified, with every muscle undeveloped, and who will suffer in her confinement from the want of physical force. There is a theory at the present day, which is a most interesting one and probably correct, advocated by Darwin and his disciples, that excessive development of the intellect from the universal higher education has produced and is still producing a relative increase in the size of the heads of the children of intellectual classes. If associated with this we have feeble muscular development on the part of the girl, what else can we expect but difficult confinements? And undoubtedly the many cases of serious womb-troubles following childbirth are due to these causes. It is not necessary, then, that exercise should mean gymnastics or a stated walk of a mile each day; but when we say that a woman while she is carrying her infant should use her muscles, we are satisfied that the ordinary exercise which falls to the lot of the housekeeper, especially if she take pride in the appearance of her own home, will be sufficient, together with the attention to those duties which call her out of doors, and give her all the fresh air and muscular development that she needs.

In the earlier stages of pregnancy,—and by early stages we mean before the seventh month,—there is very little need for a woman to change her habits as regards going about; if she keeps perfectly well, walking in moderation, driving, bathing daily, provided it is not in hot water or water that is too cold. As regards dancing, the use of the sewing-machine, swimming, horse-back-riding, these questions are constantly asked of the family physician. There are some people who will immediately take the ell if the inch is granted, and lay all the blame on their doctor; but, as it is his duty to understand all the little peculiarities of his patients, and, indeed, this study is as important and difficult as the study of medicine itself, he is the proper one to give the answers in each individual case. My own opinion is that they should be avoided; and yet a good strong healthy woman may work over a washtub doing the hardest kind of laboring work until her term is up. The answer to this is that if the young mother who reads this book is as strong as this woman who has been brought up to hardships, she probably could do the same thing.

Should our patient at any time have a sense of fatigue after exercise, it would be well to rest before meals, to loosen her clothing thoroughly, to lie first on one side, then on the other, to relieve the pressure which the strain of walking has made her child exert upon the large blood-vessels and nerve-trunks.

Indeed, towards the latter end of pregnancy, after her walk she will find a recumbent position much more comfortable and resting than a sitting one, especially if she attempt to sit in a comfortable easy-chair with soft cushions. It is very rare that a woman cannot take a brisk walk, especially in the evening before retiring; it will enable her to get a good night's rest, and at the same time she will feel a certain amount of freedom in going out at this hour without the restraint of wearing close-fitting clothes; and this leads me to speak of the way in which pregnant women should dress.

If we consider the subject of dress we shall find the object to be gained twofold,—first warmth, the other fulfilling the dictates of fashion, or, in other words, personal adornment. The matter of clothing is a very important one at this time.

Shortly after conception takes place there is a feeling of weight, which is usually experienced in the lower part of the abdomen,—a sense of fulness, which frequently gives rise to discomfort owing to the sinking in of the abdomen from enlargement of the womb. There may be, possibly, a dragging weight in the knees, backache, constant desire to urinate, and the fulness in the bowel which accompanies constipation. By the third month the abdomen will increase in size, and will gradually show enlargement; as we have before noted, the breasts will become enlarged, fuller. Then the matter of dress becomes one for further consideration.

It should be definitely borne in mind *that at no one part of the body should the clothing be so tight as to interfere with the breathing or circulation;* indeed, there should not be the least uncomfortable pressure. We may set this down as a rule to be observed in all cases. Physiologists tell us that women breathe more with the upper part of their chests than men, and that this is intended by nature to obviate the disadvantages which would otherwise occur during pregnancy. Should, then, a tight band be placed around the waist, or heavy skirts hang from the abdomen, or corsets be used to make the wearer look small and not show her condition, interference with breathing, many disorders incidental to pregnancy, would occur. The word *enceinte* itself is derived from the Latin, which signifies that the Roman women when they became pregnant divested themselves of the girdle which was always worn by married women. We have already said that the question of dress was one of warmth and adornment. As far as warmth is concerned, there is no doubt that the accumulation of fat, which is greater in women than in men, permits them to wear clothes of lighter texture than the opposite sex. The beauty and rotundity of form, the absence of angles, is due to a deposit of fat, and especially is this the case over the chest,—a delicate portion of the body, where exposure carries with it most danger.

In this latitude of 40° it is absolute folly for women, especially young girls and those about reaching the menopause, or change of life, to pass our winter and spring months without wearing woollen garments, should it only be the lightest gauze-flannel, in a complete set, with high neck and long sleeves; in fact, the argument against it is that an evening-dress necessitates the taking off of this dress, and consequently the greater exposure to cold in consequence. At the same time there is in reality less risk to be run in a heated ball-room, or when the individual is flushed with the excitement of pleasure, than when exposed to the dampness of the street, with nervous depression, in a cold wind or bleak storm, or the draughts of a house when the system is below par and the circulation feeble. All pregnant women should most certainly wear flannel in winter, or they can use the more expensive materials of wool and silk.

There is a fact which we all take for granted, though the fashion fails so far to make it a custom in this part of the world, certainly not so with our English cousins, —that all articles of clothing for women should hang from the shoulder; that is, the shoulders should bear the weight. In England the fashion is rapidly gaining ground of supporting heavy skirts by means of suspenders; at any rate, at no time should the pregnant woman have any article of clothing which will constrict the waist or drag from the loins or hips. This should be an imperative rule; and, indeed, when we come to consider the question of dress in young, growing girls, we will see why it is that heavy skirts, when fastened around the waist, are productive of harm. We may say that there are actually two waists in all women,— one just above the hip-bone and at a point where tight lacing will cause pressure on all organs contained in the pelvis (the space between the hip-bones): this is the smallest part of the body, below the last rib; the other one is made use of in high-waisted, old-fashioned dresses. The former bears the weight of heavy skirts that are belted and worn without corsets. The high- (short-) waisted dresses are certainly less harmful than low- (long-) waisted dresses. This leads to the question of corsets. Our own opinion is that a light, short corset which does not press anywhere unduly is of great comfort to a pregnant woman; it permits the skirts and undergarments to be fastened readily without compression of the waist, and in so doing distributes the weight very much as if the garments hung from the shoulder. Moreover, it gives support to the back, which is frequently restful, especially when the weight of the child becomes excessive, and when the small of the back suffers great strain from fatigue. It will also give warmth to the body,—that is to say, an even temperature, especially around the abdomen.

THE CARE OF THE BREASTS DEMANDS ATTENTION.

From the very first month we have seen that the breasts take upon themselves certain changes; they become larger, become tender, and are apt to contain, as time goes on, a small quantity of milk. The stretching of the skin from this enlargement, as also the stretching of the skin of the abdomen, causes a great deal of discomfort. The clothes should be made to fit closely without undue pressure. When the tension and feeling of fulness in the breasts and abdomen becomes intense, it can be relieved by anointing the skin thoroughly at night before retiring with vaseline, sweet oil, cocoa-butter, or almond oil. Great attention from the first should be paid to the nipple, owing to the fact that in most cases abscesses of the breast are frequently caused by an abrasion or crack in a tender nipple; and indeed at many times a depression or sinking in of the nipple will prevent the mother from nursing her child, which could be obviated by attention during pregnancy. It is well known that the nipple should stand out prominently in the centre of the breast without any indentation or gutter around it whatever. By gradually accustoming herself to it she will be enabled to manipulate it without the least feeling of tenderness. It should be sponged every morning and gently dried with a moderate amount of friction. If the nipple sinks to the level of the breast or below it, it should be pulled out, which

PREGNANCY.

can be readily done either by suction or wearing glasses made for that purpose, which can be procured at any drug-store; they are simply glass disks with a hole in the middle. In our opinion it is a great mistake to use brandy or any drying subject to harden the skin, unless ordered by the physician. The small glands which surround the nipple and are prominent in the dark areola or ring are lubricating glands, and these prevent softening or excoriation, cracks or fissures, which will eventually produce abscesses. Drying substances check the action of the glands and produce chapped, hardened tissues, which easily crack.

As regards the effect of emotions and the result of maternal impressions, from time immemorial it has been known that the closest relationship through nervous influence exists between the mother and her unborn child. To quote from Parvin, the following directions were laid down by one of the medical writers of India fourteen hundred years B.C.:

"The pregnant woman shall avoid becoming weary, ... sleeping in the daytime, watching at night, sorrow, climbing into a wagon, sitting upright, violent movements, ... and long-continued exertion. From the first day she must be cheerful, pious, and clean in clothing and person. ... She must avoid getting angry, carrying loads, or talking too loud."

The ancients, especially among the Greeks, surrounded their women during this period with all that was beautiful in art, that their sons might become strong and muscular and their daughters graceful. Undoubtedly there have been noted cases wherein maternal impressions affected the offspring, but far more frequently have been the cases where the fear of some such result has made the mother brood over sights which she has witnessed until a passing impression has indelibly fixed itself upon her nervous system and affected the nutrition of her child. A woman should avoid all emotions, should lead a life as placid as the most devout follower of Buddha, should use moderation in all things, and if perchance in the ordinary course of her daily avocations, or in the bustle and turmoil of a large city, unpleasant sights, narrations, or thoughts make impressions upon her, she should cast them aside as unworthy of consideration, and believe with the majority of the medical profession of the present day that her child will be uninfluenced by them. I cannot do better than quote a recent extract from *The Fortnightly:*

"The ideal mother is undoubtedly a woman more placid than nervous in temperament, more energetic than restless in habits, and with more strength of character and general good sense than specialized intellectual acquirements. Strong emotions, strained nerves, excitement, anxiety, absorption, are all hurtful to the unborn child. They tend to bring on premature birth; and if not this, then they create sickly offspring, whom the mother cannot nourish when they are born."

I will complete this chapter by giving a list of the things which should be provided for an ordinary case of confinement. Of course as much more can be added as the social position or wealth of the family may require. What I give here are merely the necessities:

BASKET.

Brush and comb.
Skein of white twisted embroidery-silk.
Soft fine sponge.
Bottle of white vaseline.
Sharp pair of pointed scissors.
Powder-box and puff, with talc-powder.
Pincushion.
Small and large safety-pins.
Pure Castile soap.
Pair of socks.
Some old fine linen.

Flannel or knitted band.

Flannel shirt, a petticoat and night-gown for infant.

An afghan, or piece of extra flannel, in which to wrap child.

Also piece of flannel, or old blanket, to receive it in.

In addition the baby will require eight day dresses, eight night-gowns, eight white skirts, four day flannel skirts, four night flannel skirts, four pairs of day socks, four pairs of night socks, six flannel shirts, six flannel bands, three dozen small soft linen diapers, three dozen larger cotton ones, at least two little knitted sacques.

The mother will require for herself—

One gum blanket two yards square.

One comfortable, and in addition to this an old one had better be at hand.

Some old cotton sheets.

A Davidson syringe.

A bed-pan.

A couple of binders made of well-washed unbleached muslin, not too coarse, wide enough to extend from the ribs to below the hips. In winter these can be made out of baby-flannel.

A supply of old towels or old sheets, to be thoroughly washed and packed away until required.

A glass tube for taking medicine.

Also some prepared absorbent cotton; the sublimated or borated is the best.

The use of the binder at the present day is not insisted upon by most accoucheurs, but its object is to give warmth and gentle support; that used for the child is for the same purpose, therefore if it is too tight it does far more harm than good, and had better be abandoned altogether. It certainly does help the overstretched muscles to regain their tonicity, and undoubtedly in this way allows the figure to resume its normal proportions.

A few words to the monthly nurse.

Your position is one of great responsibility, as frequently emergencies will arise that will tax your ingenuity and knowledge to the utmost. You do not degrade yourself by attending thoroughly to every detail that will add to the comfort of your patient and the maintenance of the equilibrium of the household during this trying period. You should be neat, interested in your case, guarded in your conversation, ready and willing to wait upon yourself, affable to all, have tact and dignity in your dealings with those with whom you are obliged to come in contact. Do not under any circumstances narrate the histories of other cases you have nursed, or enter into the discussion of private family affairs and scandals.

Take your patient thoroughly under your charge, and do everything for her comfort both of mind and body. When the patient is in bed, after confinement, under no circumstances allow her to be in a draught. If you open the window for a moment to ventilate or dust, always cover your patient and close the doors.

When you sponge her, surround the parts with a dry towel, so as not to wet the sheets, and only sponge one part at a time, then dry thoroughly, and do not use damp towels.

In a work of this kind it is well to give some advice upon miscarriage. By this we mean a premature birth, though in strictly medical language this name is applied only to that which occurs after quickening. There are many things which prevent the full time of pregnancy. Some of these are due to disease of the unborn child, which causes its early death; others to disease of the mother, which prevents the womb from enlarging beyond a certain point, or fevers, or severe illnesses; and others, again, are due to accidents. Frequently carelessness on the part of the mother, over-fatigue in her household duties or social requirements, ill-fitting clothing, obstinate constipation, unduly long walks, heavy lifting, or any strain or excitement. At times there seems to be a certain fatality which prevents mothers from carrying their children to full term. A woman will conceive, everything will go on well till the third or fourth month, and then miscarry; a sort of habit will be established, and she will do this for several times in succession. Her womb needs a rest, it has become irritable; and if this physiological rest is given it and she again conceives, she will go on to full term.

The earliest symptom of miscarriage is usually the appearance of a flow. If, then, immediate rest in bed be enforced,—and this rest in bed is absolutely necessary, —and proper medical treatment is instituted, it may be checked. The patient should at once go to bed, remain absolutely quiet, and send for her physician. Even after the flow has been established the child's life may be saved.

Miscarriages are by no means as simple and harmless things as some women try to impress upon themselves that they are. Many of the diseases of the womb, which entail a long course of medical treatment and give rise to permanent ill health, date from this cause, which a little care would have avoided.

Many cases of miscarriage in early married life are undoubtedly due to the fact that these young married women endeavor to conceal their condition from their friends; and, instead of adapting their life as they should to their changed state, they thoughtlessly join in the amusements and fatigues of society, dress as is most becoming to them, and conceal the annoyance and suffering which this often entails.

PART II.
INFANCY.

CHAPTER II.

CARE OF NEW-BORN INFANT.

How to cleanse a New-born Babe—The Cord, and the Care which it should receive—The Binder and its Object—Importance of noting the Child's Secretions—A Babe should be given Water occasionally—All about the Mother's Milk first making its Appearance—The Diet of the Nursing Mother—How to check an excessive flow of Milk, and to increase an Insufficient Supply—The Child's Diet—How often a Child should receive Nourishment.

It is scarcely necessary for me to dwell very long on the care of the infant immediately after its birth; that matter devolves entirely upon the doctor and the nurse. But there are a great many things that a mother should know about her infant, and these I will endeavor to explain as simply as possible. The child at birth is covered with a waxy material, differing more or less in amount in different cases. It is said that in children who have very delicate skins the amount of this material is often greater than otherwise. As it occupies every portion of the body, and can be removed by greasing the surface, the new-born child should be thoroughly anointed with either washed lard or vaseline, and then, with a soft sponge, tepid water, and a little Castile soap, given a superficial cleansing to remove as much as possible of the material. The eyes, mouth, ears, and nostrils, in fact, all of the mucous surfaces, should be carefully washed with warm water and a clean sponge or piece of old linen. This should be done after the cord has been severed. The cord, which is the attachment between the child and the after-birth, contains the blood-vessels embedded in a gelatinous material more or less thick. As soon as the child is born and takes its first breath, the circulation in the blood-vessels of the cord gradually ceases. It is not at all necessary that the cord of the child should be cut immediately after birth, nor should it be done until breathing has been fully established. The cord is usually grasped by the hands, and the jelly-like material within it squeezed with the fingers towards the mother; in this way compression of the blood-vessels takes place. A ligature is placed tightly about it, about three inches from the child's surface, and another one about three inches from that towards the maternal parts; it is then cut with scissors between the two. After the child has received its first washing the navel-string should be enveloped in a mass of the sublimated absorbent cotton, folded up in a piece of linen, laid to one side, and then the child's binder placed about it. The object of this binder is simply to give warmth, and support the cord. Undoubtedly, if the navel-string were allowed to hang, be caught in the clothing, be pulled about as the child receives its daily washing, a rupture might take place. I think undue stress has been laid upon the value of a binder to prevent rupture, and not half sufficient importance given to the binder as a protector in our changeable, treacherous climate. While the child is being washed and dressed it is well to secure deep and full respirations. It is said that a crying child at birth has stronger lungs afterwards. If it is turned over on its stomach with head lower than the rest of its body during process of washing, the mucus will not be drawn in during its inspiration, and a healthy inflation of the lungs will take place. As soon as possible the mother and child should get some sleep. After several hours' sleep the child can be again washed, if necessary, or sponged off; and this time it should be much more thoroughly done. The circulation will be more thoroughly established, and the skin will assume the delicate, soft red appearance of health. Extended experience has convinced me that for the first few days a little vaseline or cocoa-butter is better to use on the child after its bath than the dry powders heretofore in vogue. It needs but very slight greasing of the surface to make the skin soft and pliable and to prevent eruptions.

The child's clothing should have two important qualifications: it should be warm, and absolutely loose in all parts of the body.

After the child has had its first nap, attention should be paid to its secretions. If the bowels have been moved, use vaseline instead of water for its washing; if it has not passed its water, call the doctor's attention to it at his next visit: this is an important matter. It will be noted that the navel-string will come away in four or five days without any offensive odor whatsoever. This method of dressing the cord is a convenient one, as it does not require to be touched after its first arrangement. The use of the ligature is simply a matter of precaution, it is not a matter of necessity.

As a rule there is very little trouble from the navel, though possibly at times, if the child should become constipated, or have a severe cough, great straining may take place, when bulging or rupture ensues. This can be readily obviated by placing a small pad beneath the binder, not exerting great pressure; indeed, a binder which is too tight is more apt to cause a rupture than none at all. Should there be any discharge from the navel, the nurse should make a careful examination of it when the child is being washed, and if a small ulceration is found, as is sometimes the case, it can be touched with bluestone and then dressed with benzoated oxide of zinc ointment. It makes very little difference on which side the child lies after its birth. It is scarcely necessary to dwell to any great length upon the appearance of a healthy child at its birth. The soft, peach-like character of its skin, the constant tendency to sleep, which is uninterrupted and quiet, associated with an air of repose, are well-known evidences of health. It is not necessary to dwell at length on the fact that the child at this age is a noted creature of habit; indeed, as we come to consider infants as they grow, and we study the many causes for that most distressing state of affairs, sleeplessness, we can trace it back, I can almost say in one-half of the cases, to the unfortunate indiscretion in humoring the child when it is not more than a few weeks old. The young mother feels that the cry of her child must naturally come from hunger; but as nature has failed to supply material for the fountain, she is often indiscreetly advised to substitute milk and water, a little catnip tea, or Mellin's food. My own experience teaches me that this is wrong, and that the following procedure is the one to be recommended. After the mother has rested thoroughly the child should be placed to her breast. Some children will immediately proceed to work, knowing exactly what they have to do; others again may refuse absolutely to nurse. Those that seem to know all about it will find it hard at first to obtain any nourishment. After many attempts, and succeeding only in extracting what is known as colostrum, a laxative secretion which precedes the milk-flow, they will from sheer exhaustion give up the attempt and fall asleep. The infants that refuse the breast cry and fret. To these it may be well to give a linen rag moistened in water to suck, or a little water with a spoon, and then at a regular definite interval, say an hour, place them once more to the breast until they become used to it. There is no danger, under any circumstances, of a child starving to death for several hours, at least: it can exist without food, and needs only a little water; and then, if it is placed during the next twenty-four hours in daytime every hour to the breast, and after that regularly every two hours, by the time the flow of milk is thoroughly established the child will have its little programme laid down for it, from which it must not deviate under any circumstances. If it becomes absolutely necessary to feed the child, condensed milk, one teaspoonful to twelve or fifteen of water, is to be preferred. For about two or three weeks the child should be nursed every two hours day and night, and during the periods of intermission both the mother and child should sleep as much as possible. It is understood that the child should nurse at each breast alternately. This is an important matter, because a child would soon get accustomed to one breast and refuse the other, and this leads to incalculable suffering. After the child is nursed the nipple should be washed with a little plain water or soda or borax and water and greased with some vaseline.

The flow of milk generally begins on the third day. The flow may come on suddenly. There seems to be a rush of milk to the breasts. Indeed, this is usually the way the milk comes at each nursing. The woman frequently will have no evidence of milk whatever, the breasts hanging flabby and long, when, at the hour at which her child usually nurses, or even at emotion or thought of her child alone, the flow of milk will take place. This is especially the case with those who have had children before. These women sometimes make the very best wet-nurses.

The diet of a mother after childbirth usually consists of ordinary sick-diet. The object of keeping a woman on diet of this kind is simply because she is in bed, her digestive system is rather weakened by the excessive strain, mental and otherwise, which she has gone through, with fatigue, and liquid nourishment can be given, which is more readily digested and quickly absorbed than solid food. Doctors simply differ in their advice because patients differ in regard to their digestion. Sometimes a woman is ordered a mutton-chop, a bowl of thick gruel, or some milk-toast immediately after a confinement; others, again, are allowed to get up before the ninth day. The family physician knows the peculiarities of his case, and is capable of judging what is best for his patient. Strong, healthy women can do many things that the frail and delicate cannot attempt. As the supply of milk contains so large a quantity of water, almost ninety parts in a hundred, a great deal of the food which the mother takes should be liquid to supply this demand. If a mother loses her baby in childbirth, and it is necessary to check the flow of milk, not only is it customary to put belladonna-plasters on the breasts, but also to diminish the amount of fluid taken. If she wishes to increase the flow of milk, she can drink freely of milk, soups, water, also rub the breasts on the outside with some castor oil. It is probable in this way the use of alcoholic stimulants, as beer and porter in large amounts, became fashionable for the purpose of increasing the milk-supply. Though excellent at times for delicate, worn-out women, I certainly would oppose their universal use. I think the alcohol rather lessens the milk-flow. The preparations of malt are far more valuable, as they increase the appetite and aid digestion.

There being what is known as a determination of blood to the mammary glands during the secretion of milk, many substances may be carried along with it that should be thrown off in other ways; thus medicines are often secreted by the milk, and affect the child, that are intended to have a purgative action on the mother; a dose of castor oil will act sometimes in this way. Articles of diet also affect milk. We are all familiar with the garlic taste given to milk and retained in the butter, from cows turned out to grass in the springtime. And, indeed, cases have occurred of acute poisoning by milk from cows that have eaten largely of poisonous herbs in the pasture-field. Although a mixed diet, containing a free supply of vegetable food, also of animal food, of the cereals, together with a proper amount of fluid, is important for every nursing mother, there are certain times when additions should be made to one of these divisions of food to supply a demand; thus, if a child is emaciating, does not seem to increase in weight, the free use of fats by the mother, as cod-liver oil, will soon show an improvement. If there is a tendency to rickets, the mother should eat more largely of preparations containing lime, such as oatmeal, cracked wheat, or even take lime itself. A woman who is fat and well nourished may have the poorest kind of milk for her baby; all the nourishment goes to herself, the milk probably being of a very thin, watery kind. Indeed, in the choice of a wet-nurse it is often found that the lean, healthy woman gives the most and the best milk.

Constipation resulting from torpidity of the liver, or over-feeding, will frequently affect the milk and cause it to disagree with the child. Not alone is milk altered in its nourishing qualities, or in its digestibility, by materials not belonging to it or entering into its composition, but we very frequently have indigestion produced by milk which has probably been changed in its characteristics by emotion, by strong nervous impressions due to worriment, fright, or the engagements of society; certainly late dinner-parties, both from the character of the food and the late hour, would not be conducive to the production of nourishing milk. We have stated that regularity is the fundamental law to be observed by the nursing mother. Her food should be most plentiful, palatable, of the most nourishing quality, and frequently taken. Bread, butter, and milk should be used in large amounts. A cup of hot coffee, or chocolate, in the morning before rising; a substantial breakfast; the heaviest meal in the middle of the day; a light tea in the evening, with a bowl of gruel, such as oatmeal, the last thing upon retiring; a glass of milk just before or after each nursing, and possibly, if the patient feels much exhausted, malt extract will be of service.

We will repeat the same directions in regard to the bowels as are found in a previous chapter. It is necessary that they should be daily moved; this may be accomplished by a free vegetable diet, the moderate use of fruits, exercise in the open air, and an occasional enema if required. The use of purgatives should be confined to cascara cordial, compound liquorice powder, a teaspoonful at night, Husband's magnesia, or effervescing citrate. Great care should be taken by the mother that she does not take cold. It is well to have always a light shawl to throw over her while nursing her baby. Let me say a few words in regard to wet-nurses.

WET-NURSES.

It sometimes, as we all know, becomes necessary to decide on very short notice on the question of having a wet-nurse. I will not go deeply into this matter, because the circumstances of each case so group themselves as to bring the answer without consulting a book. The age of the child, the financial condition of the family, residence, are all to be taken into consideration. The wet-nurse should, if it is possible, have had other children; her child should be about the age of the one she is to nurse; she should be a perfectly healthy woman; her child should also be seen, thoroughly examined, and give all the evidences of perfect health. First of all the family record should be investigated. Did her father, mother, sisters, or brothers die of consumption, cancer, scrofula? If she has had other children, were they strong and healthy? Did they thrive upon her milk? If they died, of what did they die? The physical condition of the wet-nurse is as important, of course, as her family record. She should have no constitutional disease of any kind; her lungs should be healthy; her skin should be clean, free from any eruption, or the remains of one; her teeth should be good, which is an evidence of good digestion; her eyes bright, her muscles firm and well formed; she should have no loss of hair, sore throat, or chronic nasal catarrh; her breasts should be well formed and such as I have described on page 32. She should not menstruate. Then comes the question of temper, and indeed it is as hard here to get a satisfactory testimony as it was for the physical condition. She should be pleasant in disposition, not stupid; energetic, willing, devoid of high temper, and possessed of those traits which will soon endear the child to her. Such a nurse would probably give about two quarts of milk in twenty-four hours. Of course, much of this examination is made by the family physician, who is the one to decide whether the nurse is suitable or not, but there are times when the mother, or monthly nurse, has to decide in cases of emergency. On that account I have dwelt at some length on this subject. The nurse's child should be plump, well formed, free from any eruption. Its odor should be sweet and fresh, not sour; should it vomit, the material should be simply the overloading of the stomach and not sour milk. A healthy baby will often take more than its stomach will hold, especially if the mother's milk is plentiful and flows freely.

Having, then, decided upon the health of the wet-nurse and on the chances that her milk will be plentiful, she should be questioned carefully about her supply of milk. Indeed, the condition of her own baby will be evidence in itself. There are other matters to be taken into consideration. The nurse's child will have to be taken by some one who will give it care, so that her mind may be entirely relieved on that score, as anxiety and fretting may cause her to lose all her milk in a short time. It must be remembered that she comes from a class who, as a rule, are accustomed to much out-door life, very plain diet, and regular habits, and that a sudden change to a life which is luxury, variety in cooking, over-stimulating food, is apt to bring about a condition of biliousness, laziness, irritability of temper, which is difficult indeed to regulate. I have frequently known of the very best of wet-nurses, who have given satisfaction for a month or so, suddenly without any apparent cause lose all their milk, and the whole trouble of selection and the risk of getting a milk which does not agree with the child has to be gone over. Indeed, if under such circumstances the woman proves herself to be reliable and affectionate, it is often far better to put the child upon the bottle and keep her as ordinary nurse. When the question comes up for a decision between a wet-nurse and bottle-feeding, we should bear in mind that the child who is to be subject to city influences should be wet-nursed, especially during the hot summer months. I believe that bottle-feeding, which, as we shall see farther on, requires the greatest amount of care and watchfulness, is successful in many cases, but the more I see of it the more satisfied I am that every child should receive breast-milk until it is four months old; at least this is imperative for a city child. Of course when a child is delicate, or where there is an inherited taint in the family, such as consumption, or the family is known as a delicate one, wet-nursing becomes all the more important.

The choice of the wet-nurse, when possible, should always be left to the family physician; indeed, a doctor's examination of her milk and her baby should always be obtained. The system recently established at the nurses' directories in our large cities of having certified wet-nurses is a very valuable step forward, and should be encouraged by the public.

THE NURSING OF INFANTS.

I have elsewhere written[1] as follows, when on the subject of bottle-feeding:

How much food does a child require in twenty-four hours? So much depends on the infant; if the bowels be normal and there is no evidence of indigestion, the breath sweet, and the child seems desirous for more after it has finished its bottle, there is no reason why it should not be satisfied. A child of a month should be nursed about ten times in twenty-four hours, every two hours during the day and three hours during the night; at each nursing it should take from two to three ounces of milk. At the age of about three months it will probably nurse only about eight times, taking about six ounces at each feeding; at the end of about six months it will take about eight ounces. I believe that this would represent about the amount of breast-milk that such a child would receive.

The child gets the same food as does an adult; that is to say, the milk which forms its diet is composed of all the articles of food that enter into the diet required by a human being. These may be divided into five classes: water, casein (curds) or albuminoids, salts which go to the formation of bones and secretions, fats, and sugar, which are burned up to make animal heat, and also are valuable in nutrition. Eighty-seven parts of a child's food is water, but then we know that seventy per cent. of the human body-weight is water. Mother's milk is a bland, watery substance, sweetish to the taste, and has the property of forming curds in

---

[1] Annals of Hygiene, July 1, 1886.

flakes. This is readily noticed in children who have eaten too much, when the milk has been regurgitated; whereas the milk of the cow precipitates in heavy masses as a rule, and is on that account difficult of digestion.

A healthy infant, if properly instructed in the earlier hours of its life, will awaken with the regularity of clock-work and seek its meal every second hour. It should be placed at the breasts alternately, and after it has received its nourishment it will probably fall asleep, showing no evidence of indigestion or flatulence so frequent in bottle-fed children. Babies are very apt to get into bad habits of falling asleep after taking two or three mouthfuls. If this habit should be encouraged, it is one very difficult to break. It should be promptly awakened and made to continue the meal until it has taken the sufficient amount.

About the third month a child should nurse about every three hours, or possibly, if it is a large child, craves food and takes a great deal at once, every two hours during the day, and a longer interval at night. If the mother has a very free flow of milk, more indeed than she can possibly retain, it is well for her to wear the ordinary disk or cracker-shaped nipple-glass during the daytime, with a towel pinned over it, which will take off the surplus milk, and will prevent her from being constantly wet and catching cold.

The question is often asked, How long should a child be nursed? The universal opinion seems to be that for at least six months the child should take nothing but its mother's milk; but of course if it is feeble and weak from difficult labor, or disease, it should be kept at the breast very much longer, and should a summer intervene, as it would for a child born in October or November, no attempt should be made at weaning, unless advised by the family physician, until the following autumn. After the age of six months, all things being equal and the child being healthy, it may be well to gradually enlarge its bill of fare, in order to accustom it to a change in diet, or to prepare for any change that may become necessary. It has been said that a child digests bottle-food when it also takes breast-milk, and therefore that weaning should be a gradual process,— so gradual indeed as to take several months for its accomplishment. When it is deemed desirable to substitute breast-milk, the bottle should be given in the daytime, after the morning bath; or better still, if the nurse has the child, give it the breast in daytime that the mother may get her night's rest. By this time the child takes, as we have noted, more food at each nursing and nurses less frequently, and now it could have a bottle at about ten or eleven o'clock at night, when the mother retires; in this way she can nurse the child at the early morning hour, and thus avoid the exposure of getting up and preparing a bottle of food at that hour, if she takes charge of the child herself.

CHAPTER III.

BOTTLE-FEEDING.

The Child's Bottle and how to prepare it—Great Care in preparing the Bottle must be exercised—An Efficient Nurse indispensable to the Welfare of the Child—A study of Milk—Condensed Milk and fresh Evaporated Milk.

IF I have impressed sufficiently upon the reader the importance of care in the selection of a wet-nurse, I did not do so with the object of undervaluing the subject of bottle-feeding by contrast, although there is no question but what the valuable suggestions, the outgrowth of careful study, that have been published by such writers as Leeds, Smith, Meigs, Jacoby, and others in this country have impressed upon the community the fact that the raising of children by means of the bottle is by no means as difficult a matter as it was thought even ten years ago. They have all premised their teachings by impressing that *care* is the primary step to success. If it is a difficult matter to keep a wet-nurse in order, it is no less difficult to give the requisite attention to each bottle. *One bottle of tainted milk may be fatal to an infant*, and though a mother, or nurse, may day after day watch with the most zealous care the preparation of the baby's food, the souring of the milk, its admixture with contaminated water, the change of pasture of the cow, may bring on an attack of diarrhœa or vomiting which would be uncontrollable. I wish, therefore, to impress upon all those who have anything to do with the bottle-feeding of children, that when I state that a child that is not exposed to the dangers of a large city in the summer-time—and I make this exception—can be brought up on the bottle from the day of its birth and be free from disease, become strong and healthy, it is provided the same attention is given to it as would be given by a mother to her new-born nursing babe.

The first requisite for carrying out bottle-feeding with thoroughness is that somebody should take charge of the child who has a special interest in it. If I am talking to a young mother whose milk has given out, or whom the family physician has advised to bottle-feed her child, then she is the one to undertake the work, and either to prepare each bottle, or to superintend its preparation for a time at least. Possibly she has had a wet-nurse whose milk has gone, and it has been decided to use a bottle instead of procuring another. Then, let the nurse undertake the duties. She has a special interest in the child that has drawn its nourishment from her breast. If not, then get some middle-aged woman, not too old, or cranky, or over-burdened by previous experience, or on the contrary, a small chit of a girl who would require a nurse to look after her. Strange to say, these are often engaged as child-nurses, and no wonder the doctors are kept busy. Choose a middle-aged woman, or a strong, healthy young woman of intelligence, —one who is bright, cheerful, satisfied. Make your pecuniary arrangements with her perfectly satisfactory, so that she has nothing on her mind whatever. After you have tested her ability, give her your entire confidence; let her see that she is trusted. It is well that the child and nurse should have a room to themselves near to the mother's bedroom, and this room should have in it two things of great importance: one a small sick-room refrigerator, the other a gas-lamp or something by which the milk or water can be readily heated. It is necessary to have a nursing-bottle holding about eight ounces. A child a month old will take not quite one-half of this at each nursing. At the age of six months it should take at one feeding this bottle full.

The fresher the milk, the more readily it will be digested; indeed, I feel satisfied that the warm milk, just from the cow, is far more digestible than that which has been kept with every precaution for a few hours. There must be some change which milk undergoes, as it is noted by all observers that the milk when warm from the cow is but slightly acid, or neutral, in its action to litmus-paper, but after it has stood for a while always shows a very decided acid change. Mothers' milk is always alkaline.

The greater part of the secret of success in bottle-feeding is to have pure, fresh milk; and I would say beforehand that if there is the least doubt of the character of the milk served, there should be no hesitation about putting the child at once upon condensed milk until this matter is thoroughly arranged. Although one may be most careful in the selection of a milkman, in the city, the jolting that the milk gets in transit, the risk that is run from diseased cows, dirty cans, contaminated water in milk-houses, is by no means small, and especially during the heated season, when the child's intestinal tract is weakened. These causes of bad milk are very apt to be followed by disease; possibly this accounts for the fact that the milk from the same dairy will disagree with the child in summer that has agreed perfectly during the spring months.

So much attention has been paid to this matter recently that the public has become interested in the establishment of dairies where every precaution is taken to secure the very best of milk by legislative interference. The pasture and winter-feeding should be regulated; the health of the cattle, the methods of preserving the milk, and its transportation, looked after; the milk inspectors should be on the alert to prevent the introduction of such substances as boracic acid or salicylic acid to preserve the milk. Great care should be taken in the selection of milk, and in its preservation, even after it has reached the house, until used. The milk should be always, for a very young child, be tested with litmus; if it is alkaline, it has been made so by the addition of some preservative. Cows' milk always presents to litmus-paper more or less acid reaction, turning the litmus red. If there is the slightest suspicion that the milk is not very fresh, or that it has been subjected to much jolting, my opinion is that it should be boiled at once, and then put in a refrigerator to be warmed for each bottle. The boiling will destroy its ferments, and in that way diminish the chances for intestinal disturbances.

The question of obtaining milk from a single cow is one that has been frequently insisted upon, and if one is satisfied that such milk is obtained and is found to have agreed with the child it may have many advantages, but I think that the ordinary mixed milk from a dairy of common cattle will be less liable to daily changes; it will maintain, as it were, an average. Not only should the milk be pure and sweet, but it should be free from all matters that carry with them disease. Our medical literature contains very many authentic cases of scarlet fever, typhoid fever, and diphtheria, which were undoubtedly carried from the dairy by means of the milk, the farmer's family suffering at the time from the disease in question. If, for instance, the water of the milk-house should be contaminated by an outhouse well, and the washing of the pans convey these materials to the milk, the result of course would be apparent; or, indeed, milk undoubtedly is frequently diluted, and the water will carry the germs with it. Milk also has the propensity of absorbing odors, and gases that probably have with them the germs of disease may be absorbed by the milk and carried that way. Milk also may contain the germs of disease affecting the cow herself; so we see that there is a great risk to be run, and were we ever so careful and watchful we could only avoid the most apparent evils, and we will have to trust to Providence to save us from the others.

If each householder was more particular about his milk, gave it the strictest watching, and if the laws in regard to that outrageous and most criminal proceeding the adulteration, or diluting of milk, should be rigidly enforced, dairymen would soon feel the importance of obtaining and sustaining a reputation for honesty. It is a very difficult matter to reach the legislators of the land, those who make its laws, but possibly by placing these matters in a clear way before their wives, they will be made to see the criminality of adulteration of food when it becomes a matter of their own individual interest.

I dwell at great length upon the importance **of** step by step considering the preparation of a child's bottle, and this is done because it becomes a monotonous work, and unless the mother sees to it personally, no nurse, however devoted, may not some time or another become a little careless, and the result may be the souring of the milk, formation of the curd, and inflammation of the bowels and its consequences.

Dr. W. Thornton Parker, of Newport, Rhode Island, recommends a pure gum nipple, with two holes as far apart as possible, as the best for the nursing-bottle, and also says regarding the matter as follows: "When there is only one hole, the infant in nursing compresses the nipple and sends the milk in a stream in such a manner as often to nearly strangle itself. Milk coming through one hole is not as comfortable as when it comes through two, and the effort of nursing becomes disagreeable and wearisome to the little feeder. The best way to nurse an infant is by holding it in the arms, and give it the bottle in the same position and height as if it were really being nursed by its mother. When it has finished nursing, the bottle should be removed, emptied, and cleansed. Never should the bottle be left in the infant's care to use at will."

We all acknowledge[1] that cow's milk has the following advantages: it serves as the basis for the preparation of a milk resembling that of the human mother, it possesses all the ingredients that are necessary for nutrition, it is easy to obtain. Its disadvantages are, that the relative proportion existing between its different constituents is not that found in mother's milk, it possesses a form of casein which forms hard curds, this casein exists in larger amounts, at least twice or more than in human milk.

A certain time must elapse during which the milk undergoes possibly some alteration from exposure to

---

[1] I quote here from my paper read before the convention of State Boards of Health in May, 1886.

the air, is liable to be tainted with the germs which produce decomposition, and this indeed is the greatest objection to its use in our large cities. It is acid, though precisely what effect this has, or what it is due to, is not exactly clear to my mind.

But these objections can be readily obviated by the following means:

The milk from an ordinary dairy should be obtained as fresh as possible, mix together half a pint of this milk and half a pint of pure water, and to this should be added about two hundred grains or two heaping teaspoonfuls of milk sugar, with four grains of bicarbonate of soda; it should then be brought to a boil, after which two tablespoonfuls of cream should be stirred in, and it is ready for use, to be given by bottle or drinking-cup, at about the body temperature.

We have here a mixture which, according to Leeds, closely resembles mother's milk; we have also a preparation which has been freed by boiling from the objection stated above to cow's milk, that due to a tendency to fermentation, and indeed the milk is rendered more digestible by it.

For new-born children, or those a month or two old, we may diminish the amount of casein and increase the amount of sugar by the following means: Take one ounce of ordinary milk, three ounces of water; add one ounce of ordinary cream and about a level teaspoonful and a half (eighty grains) of milk sugar. Indeed, it is better to run the risk of making a mixture with too little casein than with too much, gradually increasing strength of the milk by diminishing the water, as the child grows older; but it should also be borne in mind that as we increase the water we should also increase the carbo-hydrates, by adding either sugar of milk or some of the malted foods. Sugar of milk rapidly sours and turns to lactic acid when dissolved in water; and, indeed, I believe that on this account there is little choice between it and cane sugar. In a case of diarrhœa, I would leave out sugar altogether. My own experience teaches me that, with care, cane sugar has not the disadvantages in most cases, in winter, that some fear.

Or we may dilute the milk as follows:

If to a *four*-ounce mixture composed of *one* ounce of ordinary milk and *three* ounces of water we add *one* ounce of ordinary cream (about fourteen and a half per cent. of butter) and about eighty grains of sugar of milk (a level teaspoonful and a half),[1] we will get a result which closely resembles woman's milk, though containing less casein and more sugar than most authorities give as the result of their investigation. Still, for very young infants this is an advantage.

Take *two* ounces of ordinary fresh milk, add *two* ounces of water. Now, add two tablespoonfuls of ordinary cream of good quality and a heaping teaspoonful (about one hundred grains) of milk sugar. Cream itself contains about three per cent. of casein. But I have insisted that there must be a certain amount of lime added to the mixture, and for this purpose lime-water can be used, a tablespoonful to the bottle replacing one of water. As I have before suggested, if there is the least doubt about the keeping of milk it should be immediately brought to a boil and then placed in the refrigerator, a certain amount being withdrawn and heated over for each bottle. Under no circumstances should a bottle of made-food be heated again,—that is to say, what remains over after the child is nursed should be thrown away. It can readily be understood why this is the case when we consider that as the child draws milk from the bottle the air which

---

[1] A silver teaspoon, such as is in ordinary use, when filled with sugar of milk and "levelled," will contain about fifty-seven grains; a plated teaspoon contains about five grains less,—practically one drachm. A "heaping" silver teaspoon holds about one hundred and seventeen grains of sugar of milk,—practically two drachms.

replaces the milk is that exhaled by the child, and acts most quickly as a putrefacient.

The milk should be given to the child about the temperature of the body or a little warmer,—that is to say, about as hot as can be borne in the mouth,—a temperature of 100°. Dr. Thornton Parker, of Newport, Rhode Island, says that the best method of preparing the milk for the bottle is as follows: For a child of three months old take of pure Alderney milk one cupful (one gill), boiling water two cupfuls (one-half pint), lime-water one tablespoonful, sugar of milk one teaspoonful. Mix carefully. Each bottle should be tasted, to see that there is nothing wrong with it and also to see that it draws well through the nipple. A black rubber nipple is certainly the best to use, and there should be a number of them, so that a clean one that has been well washed may be used each time. After a child has taken its bottle, if it is drowsy it should be laid gently on its right side and allowed to sleep. The clothes should be thoroughly loosened, and under no circumstances should it be allowed immediately after taking its food to be tossed or romped with, which unfortunately is a very common practice and always ends in indigestion.

So far we have spoken entirely of cow's milk as a substitute for that of the breast; but, as has been heretofore noted, the tendency in cow's milk is to the formation of curds that are compact and indigestible, and, though this can be to a certain extent obviated by diluting the milk as recommended, there are times when, owing to the difficulty in obtaining the pure cow's milk, which is primarily essential, or owing to the delicate digestion of the child, cow's milk seems to be indigestible. We are obliged to have recourse to some process that will render the milk more digestible, and for this purpose various means have been adopted to make up a child's bottle.

This brings us to the subject of condensed milk. A reliable brand of Borden or Canfield's has the following advantages: When diluted with from ten to five parts of water, it represents mother's milk pretty closely, depending on the age of the child, with the exception that there is less cream, but to a pint of this mixture four tablespoonfuls can readily be added. The evaporation of the milk in its preparation has destroyed its tendency to fermentation to a great extent. This most certainly is a great advantage. It will coagulate in flakes, and does not require the addition of any sugar, as by analysis it is shown that when the mixture is thus prepared the amount of sugar it contains is about equal to that in mother's milk. It can be universally obtained, and is useful on that score; its disadvantage in many instances is due to the cane sugar, and some object to it on the ground that it is supposed in many cases to lead to rickets. My own experience does not bear this out, though certainly if I were to find that a child fed on condensed milk should show undue acidity, either in its stools or its breath, due to the presence of lactic acid, I would at once change its diet. This, careful watching should avoid.

In summer weather the presence of cane sugar, which is a decided laxative, is objectionable, and herein exists the great difficulty of the proper selection of a food for that season.

In order to counteract any tendency to rickets, I usually incorporate in the milk some lime,—either limewater, or still better, I think, the lactophosphate and carbonate of lime; indeed, I would establish this as a rule in the preparation of all milk foods that require the addition of sugar.

Let us study for a moment the question of the "fresh evaporated milk," served daily in some cities by the Canfield Company, and which, I think, offers for the future the best field for infant feeding in those cities where it is daily supplied, especially in summer-time. We may add to one part of it seven parts of water previously boiled or filtered.

We find that it will be necessary to add to the half-pint of the above mixture of evaporated milk, two tablespoonfuls of cream and two heaping teaspoonfuls of sugar of milk. This will be equal to cow's milk, with about the same percentage of casein as mother's milk. The absence of cane sugar in this preparation renders it most valuable in summer in our large cities when diarrhœa is prevalent. Indeed, in such cases half an ounce of this milk in a graduated glass with four ounces of water, previously boiled and filtered, given at the temperature of the body. about 99°, without adding cream or sugar, would in many cases be a most suitable diet. If the bowels are loose, lime-water could be used. Unquestionably, disorders of the intestinal tract are produced by *fermentation* and also by *mechanical irritation* of undigested curds, and this is often due, not alone to the method of preparing the food, but also to the deficient supply of the gastric juices. If a large supply of gastric juice could be encouraged, both of these causes would cease to exist, as the acid mixture is antiputrefactive as well as digestive. We are often obliged to use some means so as to prepare the milk and destroy its ferments, and to diminish its casein, or so affect it as to allow precipitation in fine masses. The former is readily accomplished by boiling, or by subjecting the milk to heated steam, the latter by several means now in vogue:

The *first*, by rendering the milk alkaline, which retards in a measure the coagulating property of the gastric juice.

The *second*, by diluting the milk with water, which diminishes the percentage of casein.

The *third*, by thoroughly incorporating with it some material, such as gelatin, or a small amount of starchy matter, such as oatmeal-water, that will intimately incorporate itself in the casein as it falls, and thus allow the gastric juice to completely attack it.

The *fourth*, by partially predigesting the casein, peptonize it as it is called, before it enters the stomach.

We have, in addition to these, various other preparations, which are sometimes added to the milk to render it more nutritious; for example, soluble carbo-hydrates, as dextrine, glucose, or substances rich in albuminous matters. This, in fact, covers the whole ground of the various preparations used in the bottle-feeding of infants, and you will thus see that they all have some scientific basis to work upon, and their choice depends on questions of expediency and reliability, which should be studied in connection with each particular case.

Cow's milk can readily be rendered alkaline by the addition of lime-water in large amounts, soda or potash, and the curd affected thereby. I think the importance of alkalinity is somewhat over-estimated,— that is, the tendency seems to be to put too much soda in the milk; all that is required is to make it neutral, even for peptonizing purposes.

When lime-water is added to the bottle, two table-spoonfuls to a four-ounce mixture will be in most cases sufficient. It is always well to consult the physician before lime, or soda, is added to the bottle of milk; there may be reasons why a choice should be made between the two. Indeed, too much alkali may weaken the digestive organs and make the child flatulent and dyspeptic. Vichy water is a very good addition to milk instead of lime-water; if the child should continue to pass curds, it should be used in the same strength as lime-water. Dilution with water is a very important matter, because by weakening the milk with the object of diluting the curd we also diminish the fat or cream, the sugar and salts. Now, as all of these are essential to nutrition, it is obvious that by diluting them we are obliged to give the child greater bulk than it would otherwise take, and to overcome this difficulty it is necessary to add cream, sugar, and salts to the bottle in its preparation. A question of the digestion of fat is a very important one. The fats and sugars serve pretty much the same purpose in the system; they are the so-called carbo-hydrates, and go to the formation of animal heat, but the fats serve even a greater purpose: they are found essential to nutrition; they give strength, and act in that way the same as the curd or nitrogenous principle. Fat is in greatest demand at the time when animal heat is the most required, that is, during the winter months; the fats and soluble carbo-hydrates when supplied in excess are stored for future use; their excess in hot seasons is productive of intestinal disorders. In such cases a change to soups, or albuminous water, made by dissolving the white of egg in water, makes a nutritious diet and is a valuable change. The oils when stored give a condition of body which is firm and elastic to the touch, and when this reserve is called upon the emaciation is gradual. On the contrary, when the storage takes place from excess of sugar fat, the fat is not *staying* and its disappearance is sudden. This is well seen in children fattened on condensed milk to which no cream has been added. It is admirably described by Dr. S. Weir Mitchell in his book.

Lessen, then, the amount of cream and sugar for the summer season, and increase the nitrogenous elements.

The question so often arises as to the exact value of condensed milk and the cases in which it may be used, I can well be pardoned if I again dwell upon it for a few moments. The intense heat which is used in making it has destroyed the germs of putrefaction and thereby helped to preserve it. This is a very great gain. Then also the statement that only fresh, sweet milk can be condensed is undoubtedly true, as the odor which arises from stale milk would at once expose its character. The soft flocculent masses into which it is coagulated are of immense advantage, especially in young infants; it is the nearest approach to mother's milk. The only question which is at all worthy of consideration is that of the sugar which it contains, and the deficiency of cream when the mixture is diluted compared with that of mother's milk. For instance, a bottle made up of an ounce of condensed milk (mothers and nurses should use a large graduated measure in preparing babies' bottles, it is so much more reliable than the ordinary tablespoon), with ten ounces of water, is almost identical in its composition to mother's milk, possessing very much the same quality, with the exception that it has cane sugar instead of sugar of milk, and has less cream. For a very young infant, one who has been suddenly deprived of breast-milk, a mixture of this kind probably possesses greater advantages than any other milk food, and I feel satisfied that it will agree best and be more easily prepared than any other bottle. The water should be previously boiled and filtered, the can kept in a cold place, well covered, and each bottle made up fresh. I would even prefer this form of condensed milk to the evaporated fresh milk, which has no sugar, for a very young infant, unless it is previously understood that to the freshly-evaporated milk, sugar of milk should be added to each bottle.

Let me, then, be distinctly understood as recommending condensed milk, not as a regular article of diet, but simply to be used to bridge over that most delicate period in a child's existence when it is deprived, at an early age, of breast-milk, or when there are doubts as regards the character of the cow's milk from which its food is to be made. It serves as a bridge to carry the child safely over a change in the character of its food which is all-important; it also has advantages of being always at hand, and when obtained fresh and from reliable sources is usually of about the same quality. I have seen children a year or more old brought up *entirely* on condensed milk with every appearance of health and strength, and they are unusually fat children, as a rule, but at the same time I would not advise it.

Of course, as a child grows older its digestion becomes stronger; it becomes, in fact, accustomed to its food. A change can be made by adding cream to each bottle in the proportion before recommended,— that is, to a half-pint of the condensed milk as prepared above an ounce of ordinary cream can be added. If one lives in the country and milk can be obtained warm, fresh from the cow, it can be used instead of condensed milk, but I must confess that I would not recommend a city child to be given, shortly after its birth, ordinary cow's milk and water.

In addition to those recommended already, including the mixture of cream and water, which I have known to be an excellent substitute for the bottle on many occasions, we have the mixture suggested by Dr. J. F. Meigs and used by him for so many years with success.[1] A two-inches square gelatin cake is soaked for a short time in half a pint of cold water, the water is then boiled for fifteen minutes until the gelatin is thoroughly dissolved; a small teaspoonful of arrow-root, rubbed into a paste, is stirred into the boiling water, and then the milk added in the proportion of one-third milk, two-thirds water for the new-born, two-thirds milk and one-third water at six months, varying at the ages between in proportion. These are

---

[1] Dr. A. V. Meigs recommends the following way of preparing the bottle: Order from druggist a number of packages of sugar of milk, each containing seventeen and three-fourths drachms; dissolve one of these packages in a pint of water each day. Take three tablespoonfuls of this sugar-water, two tablespoonfuls of ordinary cream, one tablespoonful of milk, two tablespoonfuls of lime-water. Put in nursing-bottle, to be taken warm.

allowed to boil together for a few minutes, and then for the young infant two tablespoonfuls of cream are added to the pint of food, and to this about six and a half drachms or teaspoonfuls of sugar of milk, or three teaspoonfuls of white sugar.

Lime-water is used almost universally in the preparation of bottle-food, both with the object of making the milk slightly alkaline, and for the purpose of lessening the consistency of the curd. There are some children with whom it does not agree, undoubtedly, but as a rule it is safer to use this form of alkali than either soda or potash, because lime-water is a very weak preparation, and there is no danger of giving too much in the bottle, which certainly would be the case if soda or potash were used at the discretion of nurse or mother.

The value of whey as a substitute for milk has been advocated by many writers in cases where the child's digestion is weak and the milk curdles in large masses and very readily, or in cases of illness. Whey is certainly an admirable substitute for the ordinary bottle. It can be made by curdling the milk with rennet and straining, and according to Dr. Eustace Smith, the prominent English authority, it can be given to a child in this way:

   Two tablespoonfuls of whey,
   Two tablespoonfuls of hot water,
   One tablespoonful of fresh cream,
given in the nursing-bottle.

CHAPTER IV.

PREPARED BOTTLE.

Cow's Milk, its Advantages and Disadvantages—A Child's Digestion—Different Preparations of Bottle-Food—Why an Infant should have very little Starchy Food in its Diet—How to avoid either Constipation or Diarrhœa—A Baby's Bottle Diet may be varied—Peptonized Milk—The Care which a Mother should exercise in the Selection of a Child's Diet.

THE question of the curd commands the most serious attention. It is this curd that is always in the way, although it is an important article of diet, as it is a muscle-forming element; and yet it is not the most important by any means to the young infant, as nature has shown by supplying so little to the human milk compared with that of the cow. We must either get rid of this curd entirely for children who are suffering from disease or indigestion, or we have to so act upon it as to make it either coagulate in flocculi, or to digest it in the bottle, as has been done in the process called peptonizing.

The whey-food, or a mixture of cream and water, gives us a preparation without the curd at all, or, as in the latter, very little of it, so that children with the weakest digestions can probably live comfortably and thrive on such food; but of course when it comes to growth and development, requiring active muscular exercise, a stronger food is needed, and casein, or curd, becomes a necessity. I will quote here, for the information of those who care about the subject of digestion, a portion of a paper recently read by me.[1]

Digestion is not merely a process of disintegration; certain secretions are requisite to bring about the chemical

---

[1] Annals of Hygiene.

changes required. What are these secretions? First we have that from the salivary glands which changes starch into sugar. The saliva secreted by a child under six months is at a minimum; very little is required, simply enough to lubricate, but I may say that in a series of experiments I have recorded a child of seven days who secreted saliva which possessed sufficient diastase to convert the boiled starch used into grape sugar. This readily accounts for those infants who fatten on corn starch, much to the surprise of the family medical attendant.

As the child grows and teething begins, quite a large amount of saliva is secreted, and undoubtedly the activity of this secretion forms a prominent part in its digestive process; in other words, a child that slobbers as a rule has little digestive disturbance.

From birth the gastric juice takes a prominent part. By it the curd is precipitated and turned into peptones, or albuminose. All albuminous matter is so converted, and a burden by no means light is placed upon the liver, an organ more prominent in infancy than in adult life.

The precipitation by gastric juice of the casein presents some curious features; indeed, this matter is of fundamental importance in our studies. Woman's milk is alkaline, it is watery, its curd is precipitated in soft flakes. Cow's milk is slightly acid, its curd forms in firm hard masses of cheesy consistence. Brush, in 1879, told us that the curd in all cud-chewing animals, of which the cow represents the class, was thrown down in masses so as to be readily regurgitated by the calf for the purpose of trituration. In the non-cud-chewers the reverse is the rule. There may be other peculiarities of the curd, chemical differences, but these have not as yet been determined.

The secretion from the pancreas is the next and last of importance. It is composed principally of two materials, in fact a third may be added, the curdling principle; these will act in an alkaline or faintly acid solution: the first a material analogous to the pepsin of gastric juice, which converts casein, or other albuminous matters, into peptones, and substances that have escaped the action of the gastric juice; and a *diastase* like that of the saliva, which converts starchy matters and cane sugar into dextrine or grape sugar.

To the infant the gastric juice is the most important of its secretions; only such food as contains albuminous matter with soluble carbo-hydrates, as glucose and oil in emulsion, should be given,—such, indeed, is milk.

We have, then, two matters to consider in the artificial feeding of infants, and I shall limit myself to those within the first year: one, the preparation of a food containing the elements of mother's milk, in a combination as much like it as possible; and the other, no less important, the elaboration of those secretions which digest it. An equal balance must be maintained between the two.

The coagulation of the casein of cow's milk into hard masses can be prevented by certain means; one of the most important of them is diluting with water. It is for this purpose water is added to cow's milk; but also it has been noted that if certain materials which are not digested in the stomach are allowed to become thoroughly mixed with the milk, they will, acting in that way, so honeycomb the curd as it were, as to prevent its forming a solid lump of cheese. Lime-water may do this,—if the bottle is shaken there is a good deal of lime which is not dissolved in it,—but farinaceous food such as the cereals, the starches, if they enter the stomach as such, are not digested there, but probably act in a measure towards the curd as the sand does in the stomach of a bird. The cereals are composed of grains, when examined under the microscope, that are covered with a material that is destroyed by heat or digested by the gastric juice; the starch in either case becomes free, and the saliva, if it comes in thorough contact with it, will turn it into *dextrine* or grape sugar; in that state it is carried to the liver. The same takes place when the pancreatic secretion attacks it,—that is, after the food has left the stomach; but as a child has both the saliva and the pancreas secretion in a small amount, to feed it entirely on starchy food is simply to give it starvation diet. It cannot live on such material. Very fortunately for the baby, its corn-starch has to be boiled, and this boiling process partly converts it into grape sugar, or at least so nearly so as to allow the contact of the feeblest secretions to finish the work, and fortunately, also, nature often supplies the child with very active salivary glands during its teething period; it slobbers constantly, and the corn-starch food comes in contact with this secretion, that renders it digestible; but the poor infant who is given half-boiled arrow-root, or flour, or corn-starch too thick to flow readily through the bottle, and who cuts its teeth hard,—that is to say, has dry gums, little secretion,—will not be long before it shows an inflammation of the bowels that will be the cause of its death.

The reader can now see why it is that some children do get along well on corn-starch food and thrive from a very early age upon boiled bread and milk, cracker-dust food, or substances of that sort; but unfortunately it is these very children who form exceptions to the rule that prove the invariable evil result of attempting to give starchy food to two-thirds of the children too young. An amount of a cereal can be added to the child's bottle after it is three or four months old, if it is deemed advisable, beginning with a very small quantity.

Dr. J. Lewis Smith recommends for preparation of infants' food the following plan: Take from five to ten pounds of well-selected wheat flour, put this in a bag, tied firmly, and allow it to keep covered with water for several days, possibly a week, and this should occasionally be made to boil. In the preparation of the bottle for a child under three months, the water used for diluting the milk can have boiled with it some of this flour, grated in the proportion of two heaping teaspoonfuls to a pint; after the sixth month four teaspoonfuls.

The milk can be diluted with its bulk of water, which can be previously thoroughly boiled with either ground barley, oatmeal, or baked flour, in the proportion of a dessertspoonful to the pint, the milk poured in while the water is boiling, the whole boiled together for from twenty minutes to a half hour at least, and then strained. This can be sweetened, an ounce of cream added, and it forms an excellent food for a child after its fourth month.

If it is then understood that the addition of a cereal such as barley, oatmeal, or Graham flour is not to be given to a child as the basis of its food, but only to slightly thicken the milk and give it substance, and to prevent heavy curding, the choice of the article is a matter for consideration in each individual case. Mothers ought to know that the outer portion of the grain of wheat, corn, or oats acts as the laxative,— in other words, the bran,—and it is on that account the crushed grain is more valuable where there is a tendency to constipation. Next in order to the outer surface we have that portion of the grain in which resides most of the nitrogenous principle, the so-called *gluten*, so that in the debranned-flour we have a preparation which is nourishing and fattening, but is not laxative. The internal portion of the kernel of all these cereals contains the starch-granule, and this part we know gives the tendency towards constipation. When it is desirable to use any of these cereals it is far better to use the whole grain crushed, unless there should be looseness of the bowels or irritation of some sort, in which case the flour alone should be used. But mothers should bear in mind what I have laid stress on before, that whenever a starchy food is used the *starch-granule should be thoroughly broken up by heat, either by baking or by boiling.* This is an essential matter, and I cannot repeat it too often.

When a mother wishes to put her child on the bottle, supposing it to be about the age of four months, and we wish to add something to the bottle, it is a difficult matter to know with what to begin. Dr. J. F. Meigs advised gelatin and arrow-root; Dr. J. Lewis Smith, of New York, advocates the flour ball,—that is to say, the flour tied in a linen rag and boiled for hours, opened, and the interior taken out and grated and used with the milk. My own preference is for barley; it is the least constipating, and usually agrees well; and after the child has become accustomed to it and the digestion is in good condition, a small amount of oatmeal can be added with each, or every alternate, bottle, and a variety of diet in that way instituted.

Now, in preparing the barley for the bottle we may either take the whole-grain pearl barley, and have it crushed in a coffee-mill, or use Robinson's barley, which comes in packages, finely powdered. Of course the latter is easier manipulated and requires less time to prepare it. Of the powdered barley, take a dessert-spoonful, mix it into a smooth paste with a little water, and gradually stir this into a pint of boiling milk. If the child is under six months of age you can then add from one-half to one-third of water, and, stirring constantly, allow the mixture to boil *fully* twenty minutes. To this can be added a heaping teaspoonful of white granulated sugar and a pinch of salt. It should then be strained. Now, if this mixture is put in the refrigerator at once when it is made, in the morning, it can be used for each bottle by warming over and straining. If there is much constipation, oatmeal or Graham flour (cracked wheat) can be used in the same way. A variety of oatmeal known as the Bethlehem oatmeal comes powdered for this purpose. Of course, if the coarser grain is used, the boiling process will have to be very much prolonged; and in such cases, if simply the crushed barley, the cracked wheat, or the ordinary oatmeal is made use of, it will be necessary to put them to boil in water beforehand, say a heaping dessert-spoonful to a quart, and allowed to simmer until it is boiled down one-half. Then this can be added to the milk, stirring well, and either both boiled together for a few minutes, or, if the child is constipated, simply scalding the milk by pouring the boiling water and meal into it, stirring it meanwhile, and then strain. The sugar and salt can be added.

In city practice I always recommend the boiling of the milk for precaution sake, and I think the tendency to constipation can be overcome by adding cream to the bottle and by giving the child occasionally a bottle of water, which it will readily take. It is important to bear in mind that the food should never be made so thick that it will not run through the nipple. The food should be made in a farina-boiler, that the milk may not become scorched.

As I have said, the selection of food is to a certain extent an experiment, and therefore the child should be watched to see whether it exhibits any symptoms of indigestion. Regurgitation of food, the souring of it in the stomach, flatulence, hiccough, nausea, and finally either constipation with great pain, passage of curd, undigested milk which has a disagreeable odor, white passages, or diarrhœa, are of course all evidences of indigestion, but these must not be confounded with the symptoms that are brought about by tight bandaging, jolting, dancing the child up and down after a meal, forcing it to take more food than it can conveniently carry or digest.

I have often seen cases in which the mother had

blamed the indigestion on the diet when the food had agreed perfectly well, but the indigestion had been brought about by the way in which the child had been treated. If the child uses the bottle, it should not lie flat on its back, but should assume the same position that it occupies when nursing from its mother. It is well always to avoid the use of purgatives, or laxatives, in bottle-fed children as far as possible. Use in their stead the more laxative cereals. Remember that water is nature's most efficient laxative, that when salt is added to the food it has the same excellent effect, and that regularity and the establishment of habit, both in its feeding and the timely movement of its bowels, is of the greatest value to the child. Do not think that it is the amount of material you put into a child's stomach that is alone necessary for its sustenance. Do not think that because a child is given a quart or more of milk a day it is all that is necessary, and that it must thrive whether or not. This is a mistake which, it seems to me, mothers are constantly making. It is the amount that is digested and absorbed that nourishes, the rest decomposes or irritates.

This brings us to the subject of pre-digesting the curd.

Professor Albert R. Leeds, of Hoboken, gives us the following way of preparing milk for infants:

> One gill of cow's milk, fresh, unskimmed,
> One gill of water,
> Two tablespoonfuls of rich cream.

To these can be added one powder which contains *two hundred grains of sugar of milk* and *four grains bicarbonate of soda*, with *a grain and a half of extract of pancreas*. These powders, each one containing the above formula, can be made up in any drug-store. The milk, with this powder added to it, should be put in a nursing-bottle and placed in hot water,—water so hot that it cannot bear the hand more than a minute at a time,—and kept there for about twenty minutes, and allowed to cool sufficiently for the child to take. This powder is called *Peptogenic Milk Powder.*

In a lecture before the Philadelphia Hospital Nurses Training School, Mr. Fairchild spoke as follows:[1]

"When we speak of peptonizing food, we do not mean that pepsin is employed in the digestion of the food. We simply mean that the albuminous portion, the casein of milk, for instance, is converted into peptone. The materials used for the purpose are the ferments of the pancreatic juice in the form of a powder, —the extract of pancreas.[2] This contains several ferments, each of which acts on a different form of food. The conditions under which these ferments act are very simple. If, in attempting to digest milk, you add the ferment to the milk when it is very cold, no action will be obtained. Again, if it is added to very hot milk, no action will be obtained. The conditions under which these ferments act are similar to those found in the body. A good test for determining the proper temperature of the food is to taste it. If it is hotter than can be sipped with ease, it is too hot. If it is desired, a little thermometer may be employed to obtain the proper temperature, which is 100° F.

"I shall now show you how to make peptonized milk. I take one of these 'peptonizing tubes,' which contains five grains of pancreatic extract and fifteen grains of bicarbonate of soda, and empty its contents into a quart bottle. To this I add a gill—that is, four ounces or eight tablespoonfuls—of cold water, and if it is for infant feeding, the water had better be previously boiled. Enough may be boiled in the morning to last all day. I next pour into the bottle a pint of milk, and shake the bottle well. By adding the water and

---

[1] Reported by Dr. William H. Morrison, Holmesburg, Pa.
[2] Or the "essence of pancreas," or the "liquor pancreaticus."

milk cold, we run no risks. Having done this, the bottle is set into a bowl of warm water, which should be of such a heat that you can hold your hand in it for a minute; the temperature of the milk is thus raised to about 100°. The milk is allowed to remain in the water for half an hour; it is then put upon the ice, and the digestion will still continue for some time,— until the milk is thoroughly chilled, after which no further digestive change can take place.

"I have here a bottle of milk which has been digested in this way. I have allowed the digestion to be carried a little further than is usually necessary, in order that I might show you the properties of peptonized milk. I first take ordinary milk and add an acid to it. At once a mass of coagulated casein falls to the bottom of the glass. I treat a sample of peptonized milk in the same way, and there is no trace of casein. As I have said, it is not usually necessary to digest all the casein, and the directions which I have given are for average cases. You have to observe the effect of the milk on the patient's digestion. If it is assimilated readily, the proper pre-digestion has been secured; but if it is necessary to digest it still further, the milk may be allowed to remain longer in the hot water. If the milk has been digested too much, and is a little bitter, it may be made agreeable by the addition of a little sugar. You will soon learn by experience how to adapt the process to the requirements of each case.

"Instead of using plain water, we may take a quantity of starch paste, add a little pancreatic extract to it, and stir it up. When starch is boiled it simply swells up, but within a few minutes after the addition of the pancreatic extract it becomes a thin liquid from the digestion of the starch,—by the pancreatic diastase; this starch is now in the way of being converted into glucose or grape sugar. This may be put into a bottle with the soda and milk, and digested in the manner just shown, and we shall have peptonized milk gruel. Here the nutritious elements of the starch are added to the milk. In using the peptonizing powder a little water is always used to dilute the milk, otherwise it would be slightly curdled by the extract of pancreas.

"This peptonized milk can readily be made into lemonade. It may strike you as rather odd to add lemon-juice to milk, but as the milk has been completely peptonized, it will not curdle, and the lemon is often desirable to make the milk pleasant. Rum and sugar may be added if stimulants are required, making a delicious punch. It may also be taken with carbonic acid water, and, if thought necessary, lime-water may be added, although we have already added an alkali.

"You may make a peptonized milk jelly. If you wish to make a jelly, it is necessary to allow the digestion to progress for a longer time. A pint of peptonized milk is heated to the boiling-point,—that is, you scald the milk. This is necessary to destroy the ferment. Then take three-fourths of an ounce of Coxe's gelatin, a tablespoonful of lemon-juice, and a couple of tablespoonfuls of orange-juice. When the milk is scalded some of the lemon and orange-peel may be scalded with it, which gives a fresh flavor of the peel. The gelatin is then added, and wine, brandy, or St. Croix rum may also be added. If you do not remember to scald the milk, you will not get a jelly, for the extract of pancreas will not only digest the casein, but it will also digest the gelatin.

"In digesting meat, take two tablespoonfuls of chicken or beef finely minced and boil it with a gill of water. This makes the meat soft and facilitates its digestion. The meat is then rubbed into a fine pulp and put back into the water. You may now add a gill of the starch mucilage and one of the peptonizing powders. It is then set aside for two or three hours, and at the end of that time scalded. The peptonized soup may be seasoned to suit taste. The scalding is necessary to stop the digestion, which otherwise would go on and lead to but refractive changes.

"This plan may be used with ordinary soup. Take two or three tablespoonfuls of the meat, barley, etc., strained from soup, rub it to a pulp, and add fifteen or twenty grains of pancreatic extract and half a drachm of bicarbonate of soda; add to a pint of the soup, and proceed as just shown. There is no doubt that you get artificial digestion of all the substances, and at the same time you have no more trouble than in making ordinary food. If this is strained, and gelatin added, you obtain a nice clear jelly. The peptonized milk jelly is more agreeable than those made with ordinary milk.

"In preparing peptonized milk for babies, we follow a little different plan. In using cow's milk we have to dilute it with an equal quantity of water in order to obtain the proper amount of casein. We have to add a small quantity of milk sugar to make up for that lost by the dilution with water, because mother's milk contains a little more sugar than cow's milk. Then we have to add the alkaline salts which are found in human milk. Dr. Keating spoke of the acidity of cow's milk, and this is a point which few people properly appreciate. Testing this sample of milk with litmus-paper it is found to be distinctly acid, and, in fact, I have never tested cow's milk without finding this acid reaction. Here we have a powder (peptogenic milk powder) which presents the proper proportion of milk salts, milk sugar, and the digestive ferment to change the casein into the soluble form in which the albuminoids exist in mother's milk. I take four ounces of milk, add the proper amount of the peptogenic milk powder; next we add four ounces of water and two tablespoonfuls of cream. This latter is an important element, for mother's milk contains more fat than cow's milk. In this way we obtain the same *proportions* of the different elements as are found in human milk. All that is now required is to heat it to the proper temperature for five or six minutes in order to properly modify the casein. The temperature is to be determined as in the former case by sipping or by the use of the thermometer. In this process, having first made a milk mixture which contains the right quantity of all the elements of mother's milk, and with its peculiar alkaline character, then we seek to effect just such a change of the casein—the 'curd'—as will present it to the infant's stomach in the condition fit for digestion, in such a condition that it will behave in the stomach just as mother's milk does, and make the same demand upon the natural digestive functions. If, however, the baby is very ill, and not even capable of digesting its natural food, this method allows you to digest it still further.

"There are two other ways in which the 'humanized' milk may be prepared. Instead of taking the quantity of milk which I have done, we may take a larger quantity and a proportionately larger quantity of the other ingredients, mix them, and keep them on ice. There will be no action as long as the low temperature is continued. The proper amount may be poured out and heated whenever it is required.

"The other way is to make the mixture as just described, and stand it in warm water for fifteen minutes. This will give the proper amount of digestion. Then scald it; this kills the ferment, and the milk can be kept with no more care than ordinary milk. You can then take the proper amount and warm when it is needed, and you have no further trouble with it. For asylums where there are many children, this is probably the best way. This gives us milk which is as exact an imitation of the natural milk of the mother as we can expect to obtain it in practice."

The extract of pancreas can be obtained at any drugstore, and is at present highly recommended by all physicians in this country and Europe when prepared according to the directions just given, for infants who are suddenly deprived of mother's milk or for those that are sick. A certain amount of care is required in the preparation of this food, because if the peptonizing process goes on too long the milk will become bitter and the child will refuse it; if it does not go on long enough, the curd of course will not be affected. I think on this account it is considered troublesome, and, in institutions especially, condensed milk appears to have the greater number of advantages. There is another quality which the extract of pancreas possesses which is as important as that of the digestion of casein, which is that a small quantity of it when added to the broken starch-granule will aid in converting it into grape sugar and thereby render it digestible; for instance, if a child's bottle be made up of barley-water and milk or oatmeal, a few drops of the extract of pancreas in solution will render certain its digestibility. And for children who have a tendency to diarrhœa, or with whom starchy food fails to agree, this can be made use of.

This brings us to another subject. Anything which will convert the cereals into grape sugar before the food is taken to the child, will aid in nutrition. Why? Because these cereals not only contain starch, which goes to supply fat and heat, but they also contain albuminoids, as gluten and other nitrogenous materials, which go to the formation of muscular tissue, and salts, which are bone-forming. If the whole grain can be so prepared as to be perfectly digested, a great deal will have been gained in the nutrition of the child, and for this object various foods have been suggested. A substance which converts starch into sugar is diastase, or malt. Each granule of the cereals possesses a certain ferment which, if allowed to develop by heat and moisture, will turn the starch into sugar. This is made use of in the preparation of food for children, such as Liebig's foods, where the starch has been turned into grape sugar by malting. Mellin's food and Horlick's food are prepared on the same principle. These foods are nothing more than malted grain. It is a fat-making and nourishing food, which when taken into the stomach will increase the nutrition of the body and store up a certain amount of fat, and valuable because it requires very little digestion. The child at birth, however, requires simply milk of the character which nature presents. It needs no more albumen and no more sugar. What we want is to supply a milk as nearly as possible of the quality furnished by nature. It is therefore, in my opinion, not necessary to add Mellin's or Horlick's food to the milk of children at birth. All that we need is a milk which will be digested and readily absorbed.

Any preparation of malt that will aid the digestion of starchy food is useful, not only for adults who suffer from flatulence and debility of the digestive organs, but it is also good for infants, when given in moderation; and when I advise, as a rule, mothers to avoid these preparations for their very young babies, I am only speaking of healthy children. This leads me to the question of those foods in addition to the child's bottle which either aid in the digestibility or are themselves of value in supplying nourishment. According to Professor Leeds, of Hoboken, these classes of foods may be divided into the *milk foods*, the *farinaceous foods*, and the *Liebig foods*. I give many of the different preparations under these headings, in order that the mother can intelligently make her choice, should one not agree with her child.

I have said before, and I may repeat it here, as it is a very important matter, that the choice of a food for a child is a matter of experiment, for what agrees with one may not agree with another, even in the same family ; that the test of whether or not a food agrees with a child is if the child thrives upon it,—if it sleeps well, its flesh becomes firm, its digestion is good, its temper is amiable,—because a cross child, in nine cases out of ten, is either a dyspeptic, or a sickly one. These tests are the only ones that are of value, notwithstanding the advertisement that *such* and *such* foods are the only ones that agree with the baby. Should a child be languid, drooping, appear weak, and not thrive, one of the milk foods—such as Nestle's, pre-eminently—would be a useful addition to its food, or it can be used simply with water, as it contains condensed milk, also Carnrick's Soluble Food can be advised in this way. As the child grows older, a farinaceous food may be given in the way described on page 71. Should the passages become constipated and there be much flatulence, the child suffer from colic, become restless at night, lose its appetite, then the change should be made to one of the Liebig foods, and in this way its digestion encouraged and nutrition established. An intelligent mother watching carefully her child can thus be guided in the choice of its food; but it should be always borne in mind that as milk contains all of the elements for nutrition in such proportion as is required, those foods which are not *milk foods* should always be made up with milk in the preparation of the bottle; and if fresh cow's milk cannot be obtained for this purpose, a *milk food* well diluted should be used, such as ordinary condensed milk, Borden's or Canfield's *fresh evaporated milk*, or one of the *milk foods* given in the table:

MILK FOODS.

Nestle's,
Carnrick's Soluble Food,
Anglo-Swiss,
Gerber's,
American Swiss, and others.

FARINACEOUS FOODS.

Blair's Wheat,
Hubbell's Wheat,
Imperial Granum,
Hard's Food,
Ridge's Food,
Robinson's Patent Barley,
Bethlehem Oatmeal, and others.

LIEBIG FOODS.

Mellin's,
Horlick's,
Lactated Food,
Hawley's,
Keasby & Mattison's,
Savory & Moore, and others.

Frequently a child may be so weak or exhausted from disease or from inanition that food of the mildest character will not remain on its stomach. It would be useless to keep on diluting condensed milk, as it would render it valueless. In cases of this sort, the white of an egg shaken up in a bottle of warm water to which a couple of grains of lactopeptine is added, sweetened and given by the bottle if the child will nurse, and by spoon in small amounts if the child will not, is very nourishing. Wine-whey can be given in the same manner. Gum-arabic water will nourish for a surprisingly long time, and allay irritability of the stomach and bowel, and finally the child can be gradually encouraged to take small and repeated quantities of peptonized milk or one of the *milk foods.*

CHAPTER V.

WEANING.

Weaning—When to wean a Child—Where to wean a Child; and how to wean it.

The question is often asked, At what age should weaning begin provided that there is no immediate necessity, and how should the process be managed? It greatly depends on the family arrangements for spending that season of the year which in this part of the world is most to be dreaded, the summer. Of course, if a child is to be taken to the sea-shore, or some cool summer resort, where milk can be supplied fresh and abundant, the question of weaning in the summer-time has not half the importance attached to it as it has to those who are obliged to spend the summer in, or near, one of our large cities. My own opinion is that if a child has been nursed for four months, certainly for six months, the gradual addition to its dietary of a bottle will be of advantage. I have studied this matter considerably of later years, and somewhat modified my former views on the subject, and my own investigation and those of others have recently confirmed them, that a child, partly nursed and partly bottle-fed, after its fourth month, thrives better than one bottle-fed alone, —that, in fact, breast-milk helps to make the bottle more digestible. Not only is this the case, but I believe that it is a great relief to the mother, gives her more time to rest, is less drain upon her nutrition, and it also is of importance should the child be obliged to take the bottle, either because the mother's milk gives out, she becomes pregnant, or her health gives way; but I certainly think that four months is young enough, provided the mother's milk is found to be of good quality, nourishing, and the child thrive upon it. Should this not be the case, the addition to nursing by an occasional bottle will have a tendency to concentrate the breast-milk, making it more nourishing, and thus avoid the necessity of weaning altogether.

For a child that is born as late as January or February I should not recommend the addition of any bottle food until the following October, provided the mother is able to nurse it. For a child born in October or November, and especially when the following summer can be spent out of town or at the sea-shore, the weaning process could possibly be all over by May. If a mother is strong and hearty, has no consumption in the family, has plenty of milk, and is not in the least pulled down by nourishing her baby, if she can nurse it a year I think it so much the better, and for the last four months her child can take some bottle food in addition. I do not think any mother should nurse her child for more than a year; there is no necessity for it as a rule. The milk is not sufficiently nourishing, and unless it is supplemented by the bottle the chances are that the child will become sickly. Indeed, a good strong healthy child that becomes accustomed to the bottle food will wean itself before that time is up, and this is exactly what we wish a child to do,—to wean itself. Now, in the gradual process of weaning or the addition of a bottle to its regular nursing, the babe of five or six months may take its food according to the following programme: It should nurse from its mother in the morning, six or seven, and then after its bath take a bottle about half-past ten or eleven o'clock; possibly nurse from its mother about one or two o'clock; again a bottle at five or six; and then the mother should nurse it at retiring at ten or eleven. In this way it becomes gradually accustomed to the bottle at the time of day when it is most apt to agree. It gives the mother an opportunity to take exercise and rest. Possibly it may need a bottle about four o'clock in the morning.

The question would be asked, What should be the first choice when selecting the bottle food of a child at this age? I would recommend that the bottle be prepared with barley, or Imperial Granum, cracker-dust, or grated flour ball, and add in increasing quantities, beginning probably with a small teaspoonful, Mellin's or Horlick's food. As the child grows older, the midday bottle can be varied. I believe that by using a mixture of the cereals we often get a more palatable and more nutritious preparation than by using them singly; thus, oatmeal and barley, or Graham flour, can be used together, or oatmeal and Hubbell's food. The child will show a decided preference for some kind of bottle food, but bear in mind that it is a great mistake to stuff a child. A mother will often be tempted to add a little more oatmeal, or Mellin's food, etc., thinking that a little increase will make her baby stronger and rosier. This is true if the child lives in the country or at the sea-shore, where it is out of doors all day long, in a cool bracing climate, especially if it is able to run about; but if mothers could see, as doctors do, the numbers of sluggish, heavy children that our cities afford, with coated tongues and heavy breaths, and constipation, who are fairly packed with baby foods and all the most concentrated articles of diet that modern chemistry presents us with, they would understand why it is that the ill-fed, ill-clothed children of the poor, who live on a crust and digest it, are so much more able to resist disease than their own. It is on this account I believe that although oatmeal is a most valuable addition to the diet, it should only be used in small amounts, should be thoroughly boiled, the children when taking it kept out of doors as much as possible, and it should not be used in hot weather, nor with children who have what is called a " bilious tendency."

If fresh cow's milk is not obtainable, condensed milk can be used in this gradual weaning of children. Let me here say again that when we use any substance besides milk in its bottle, such as the cereals, we should not forget that the child needs water—pure, clean water to drink. Very often a child that is partly bottle fed and partly nursed is restless at night, will not sleep. Instead of the mother's trying to put it to sleep by nursing it, if she would simply give it some water in its bottle, or possibly a little Mellin's food in the water, it would go to sleep and not run the risk of indigestion from over-feeding.

Bear in mind that in the summer-time if a child takes more food than it can digest, this food is apt to decompose, act as an irritant, and possibly give rise to an inflammation that will end in summer complaint. Suppose, then, that a child has been weaned from the breast, and the object now is to gradually take it off the bottle or give it some additional food besides that which it takes in its bottle. We may presume that its bottle has been agreeing with it, but that for the last few weeks it has turned against it, as it were,—it seems to crave more solid food. Certainly by the time it is twelve months old it could very well be given a small cupful of chicken-broth or beef-soup. Possibly before this time it has been given, instead of the usual bottle after waking from its mid-day nap, some boiled bread and milk; and now, instead of bread and milk, some chicken-broth, with a little dry toast soaked in it, can take the place of this meal, and the bread and milk be given for supper about six o'clock. In this way the "bill of fare" for three meals can be gradually mapped out, and the child permitted to masticate part of its food; this will aid the cutting of the teeth as well as increase its digestion. When a child takes bread and milk in this way it is always well to let it have an occasional drink of plain milk.

CHAPTER VI.

FRESH AIR, VENTILATION, OUT-DOOR EXERCISE.

The Great Importance of Ventilation—The Selection and Care of the Nursery—How it should be Heated—The Danger of allowing Children to be left to the Care of Young and Inexperienced Girls—How Children frequently contract Diseases.

THERE are a great many facts that have been given us by scientific men that should be applied to every-day custom; among these none more important than those relating to the subject of ventilation, by which we mean the getting rid of foul air and the entrance of pure air in as easy a manner as possible, free from draughts. We all know that the passage of cold air through a chink, when striking against some sensitive nerve-point on the surface of the body, which is exposed, will have a peculiar way of abstracting heat, or give what is known as "a cold," affecting the mucous membranes. Of course, the more delicate or the younger the individual the more susceptible it would be to such an impression. There are certain parts of the body more susceptible to these currents of cold air; these are the face, neck, and feet; neuralgia, sore throats, and colds in the head being the consequence. But it is to be observed that these draughts are more apt to make themselves felt when the parts on which they strike are in a state of relaxation; naturally, should there be perspiration, its evaporation would intensify the cold impression. When children, then, after active play, perspiring freely, sit in a room in a draught, they will take cold, and at the same time they might continue their play in a colder room, and not feel it; the action of their muscles, the excitement, will give them resistance which they would not have in a state of quiet. It should be our object, then, in the choice of a nursery, to have a room, or two rooms communicating (when speaking of these matters I specially refer to our city houses), as far removed as possible from the contaminations and the filth of the streets. The play-room should be large, certainly, situated in the second or third story, and should have the sun in it at least part of the day.

Of course, for a few weeks the new-born babe will sleep with its mother; if it is a strong and vigorous child it can be placed in a crib or bassinet by her side. The great objection to placing the child away from the mother is that, after the nurse leaves, she would have to reach for it, and if her room is cooler at night, as it should always be, there will be great danger of her taking cold when she nurses the baby. On this account, the child's crib should be placed as near as possible to the mother's bed, or she should manage to allow it to sleep in the bed with her, but so arranged that it will have a portion of the bed entirely to itself. The child is, certainly, after a couple of months, healthier when sleeping alone.

For the first month, at least, after birth, while the nurse is still with the mother, she should take charge of the infant and bring it in to nurse at the proper hour. The choice of a nursery and sleeping-apartment is a matter of great importance. The essentials are, purity of atmosphere and uniformity of temperature, freedom from dust and gases which may arise from methods of heating; especially is this the case with faulty hot-air furnaces. There should be plenty of sunshine when possible. At the present day, with our extended knowledge of the cause of disease, impurity in the atmosphere has assumed great importance; we are now able to recognize the fact that certain diseases which we hitherto attributed to cold are in reality due to filth; that certain other intestinal disturbances, which were attributed to heat, are in reality due to decomposed or fermented food; these facts are most important for us to bear in mind, not only to enable us to cure disease, but also because by a thorough recognition of them, these diseases or disorders which have been attributed to climacteric disturbances may, by the timely institution of hygienic measures, be avoided. There is no reason why a child that has a well-ventilated, clean, bright nursery, whose milk is watched with zealous care and never allowed to be tainted, one who is daily bathed, not overfed, neither debilitated by too heavy clothing nor subjected to daily fluctuations in bodily temperature, should not pass through the dreaded summer season in a city unharmed by the so much dreaded summer weather. I mention this because, probably, many mothers who read this book cannot afford to spend the summers out of town.

I think a nursery should always have an open fireplace for ventilation, and a counter-opening should be made over the window so as to allow the air to be changed with the least draught possible; this can be done by pulling down the top sash, putting about a four-inch strip of board to keep it down, and an indirect current will be made between the sashes. Of all the methods of heating, probably the most scientific, but unfortunately the one that can be most abused, is the hot-air furnace. Parents should see that their nurseries are supplied with air that is, first of all, pure,— it must be taken at as great a distance as possible from the ground, not immediately off of gutters and damp yards, as is usually the case. The air most charged with atmospheric impurities, whether they be germs or gases, is that which is usually heated and sent to the nursery; the heating simply making it more poisonous than before. The air should be taken as far from the ground as possible, heated, and then passed over a surface of pure water; it will receive a certain amount of moisture, and then be carried, as free from dust as possible, directly to the nursery. Air which is not passed over water but simply dried will, undoubtedly, produce various forms of irritation of the mucous membranes, dryness of the throat and nose, languor,—symptoms which we all recognize at once. Certainly nothing could be better than an open grate, with a wood-fire, even if it be only occasionally used, to supplement the furnace, especially at night.

The sleeping-room should be heated through the nursery, if possible, and should be only occupied at night. Of course, I recognize that these matters are difficult to control, but at the same time, if parents know what ought to be done and take an interest in the matter, think for themselves, many arrangements can be made to overcome temporary difficulties which at first seem unsurmountable, and render a dreary, unhealthy nursery, healthful and habitable. The nursery should always have a thermometer, and the temperature be kept about 70°. The heat should be always shut off at night, and the child's clothing so arranged as to prevent its being thrown off while the child is perspiring during the early morning hours, when the system is most depressed, at which time the danger of catching cold is most imminent. Avoid all sewerage arrangements,—pipes of every kind,—in a nursery. The science of ventilation and house-drainage gives us probably as perfect a system as we shall ever have, but, unfortunately, the slightest fault of construction will turn the otherwise harmless contrivance for our comfort into one of the most deadly; and there is no means by which we can detect the presence of the sewer air that serves as a carrier of the poison of diphtheria or typhoid fever, any more than there is evidence of its existence in drinking-water, which equally will serve to disseminate this poison. The most costly habitation of the wealthy will find these dreaded diseases carried into its midst to a greater extent even than is found among the poor. The classes of persons who suffer most from diphtheria are the very wealthy and the very poor. Those of moderate wealth guard their children carefully against cold, in the first place, and their means prevent them from having the luxuries which carry the deadly sewer air into their bedrooms. Physicians believe that diphtheria most frequently requires a cold, a catarrh in the throat, before the poison is thoroughly absorbed, and probably this accounts for the resistance which is observed in so many cases against an attack in the summer. Just as soon as the child takes cold, becomes a little run down, that dreaded disease, diphtheritic croup, will show itself.

I have laid great stress on the importance of thorough ventilation and fresh air, but I wish to be distinctly understood that constancy in the purity of air, both day and night, is not only requisite in the nursery, but also in the sick-room. Cold air is not necessarily pure air, nor is air which is warmed made impure by warming; at the same time it should be remembered that frosty air, filled with germs of disease, may be harmless to breathe, but the same air warmed by a " heater" may become most deadly. A child's vitality is lowered at night, its circulation less, its resistance to disease less. A person sleeping and chilled is much more liable to take cold than one who is awake; especially is this so in childhood and old age; but the temperature of the sleeping-room may become reduced at night, when the heat is turned off, with benefit, if the child is covered —not enough to induce perspiration, and the clothing so arranged as not to be thrown off. Of course a child, up to at least six months, should be kept in a room which has as nearly as possible the same temperature day and night, as it sleeps most of the time, and when taken out of doors is so warmly clad that the change of air cannot affect it.

To sum up, then, a nursery, or child's living-room and bedroom, should be kept scrupulously clean, thoroughly aired, freed from dust; dust, independent of its irritating character upon the mucous membrane, is the means of conveying disease. There are many days in our treacherous climate when a child cannot be taken out of doors; indeed, there are many days when a child had better remain in its well-aired nursery, days during which, if it went out of doors in its perambulator, it would inhale the exhalations from the foul masses that accumulate in our city streets. A child in arms is far safer when carried out of doors for fresh air than one in its perambulator, or one who walks.

The day nursery should be supplied with plants. They are undoubtedly beneficial to health when properly cared for, and make the living-room bright and cheerful, and this reacts on the disposition of the child. A bright, happy home makes a bright, happy child, and what is taken for temper, perverseness, in many children is often sickness and unhappiness. Those who are much thrown with children recognize this fact; indeed, it is one which every grown person feels when he or she looks back to childhood days. The energy, buoyancy, which comes with good health, is in marked contrast to the depression and irritability that is associated with illness, or, if not exactly illness, with those sedentary pursuits that are in themselves unhealthful.

The question is often asked at what age a child should go out of doors, and whether it should go out every day, notwithstanding the weather. It makes a very great difference whether the parents live in the country, or the city. Country children, of course, are out most of the time, as they grow older especially; whereas, in the city, the impossibility of thus turning them loose and the necessity of a nurse to accompany them are matters, of course, that have to be taken into consideration. After a babe is about six weeks or two months old, if the weather is at all moderate, the nurse can wrap the child well and take it in her arms out for a walk. There is less risk of young children taking cold than older ones, from the fact that they are much easier wrapped and kept warm, and the nurse is able to carry them. At the same time, if the house is well ventilated and warm and the weather cold and changeable,—dirty streets, snow, and dampness rising from the ground,—it is far better for the child to remain in the house. As soon as a child arrives at that age when it is a drag upon the nurse, difficult to carry, and at the same time cannot walk, and a perambulator is required, the time to exercise the most judgment has come. Any mother can see this for herself by going to one of our city parks and watching the congregation of nurse-girls assembled, noting the position of the baby carriages and the condition of their occupants. A child will be left facing the bleakest March wind, or the midsummer sun fiercely attacking its unprotected head, while the nurse will be engaged in conversation with a number of her friends. I have often been at a loss to know how mothers could select these young, inexperienced creatures to take care of their children, knowing full well what would be the consequences, and then be surprised if the child should be taken with a severe sore throat, earache, pneumonia, or inflammation of the brain, as a consequence. It would be far better if all children, until they were old enough to sit up by themselves, were carried by their nurses in their every-day outing, and that after a child was too big to carry, and too young to walk, it should sit up in its carriage well wrapped, then the nurse take a long walk, with the distinct understanding that under no circumstances is the carriage to be stopped; when she is tired she is to come home. I am very particular in laying stress upon this matter, because, notwithstanding all that has been written on the subject and the full knowledge that mothers obtain from their family physicians, who are one and all opposed to the present system and acknowledge that a larger part of the diseases of children is due, undoubtedly, to the carelessness, in one way or another, of their nurses,—these girls, without any experience whatever, with no judgment, certainly no affection for their charges, hired and intrusted with the care of an infant, and allowed to take it out, going where they will, carry it into heated rooms, leave its out-door wraps on, carry it out of doors while it is perspiring, expose it to contagion of every kind, taking it into all sorts of atmospheres; and yet after the child has been returned, its fond mother will fondle and caress it, guard it against the least exposure, treat it as the tenderest flower, be struck with wonder and surprise when it is taken ill. So important do I deem it that a child's nurse should be selected with the greatest possible care, that she should be a woman chosen on account of her experience, conscientiousness, and truthfulness, that I believe the mortality from contagious diseases, and from those disorders due to direct exposure, would be diminished if mothers could be made to appreciate this matter.

CHAPTER VII.

BATHING.

The Importance of Bathing—A Child's Time for Bathing must be regulated by the Child's Condition, etc.—The Cruelty of forcing Young Children to have a Plunge in the Cold Sea—A Child should love its Bath, and how it can be taught to love it.

IN speaking of the new-born babe, it was noted that immediately after birth, instead of being washed, it should thoroughly be greased with some material. The best substance to be used for this would probably be washed lard, or the non-scented white vaseline. The child can then be gently washed in warm water with a soft sponge, or soft linen, with Pears' non-scented, or Castile soap, and care should be taken that every part of the body be carefully washed, so as to free it from any impurities that may have secreted and caused irritation to the tender skin. It is not necessary that at first the child should receive a thorough washing, but certainly by the end of twenty-four hours its skin should be soft and pink, and every particle of secretion be removed from it; the nostrils, the eyes, the mouth, the various crevices of the groin and the arm-pits thoroughly cleansed.

There should be a thermometer always in the nursery, and the child's bath should be always regulated by it. The temperature of the water should be 95°, and as the child grows older and becomes strong, the circulation well established, the temperature should be gradually reduced until it is about 75° or cooler. It is not at all necessary that a child should receive a full bath twice a day; once a day is amply sufficient, in the morning. At bedtime a sponging off will be enough, unless the child is one of those excitable dispositions to whom we have already alluded, the child of intellectual parents, or those who depend upon their brains for a livelihood: these children always exhibit more or less of a nervous, irritable disposition, which renders them at times restless and sleepless. For such, the sedative effect of a bath at night is most marked; indeed, for these it is well to usually sponge off with cool water in the morning, and leave the bath for the night, making the water about 95° in order to get its full sedative influence.

It is a very great mistake for a child to get accustomed to bromide, valerianate of ammonia, brandy, or gin, to make it sleep; these should never be used without the consent of the physician; but the sedative influence of a warm bath, or warm foot-bath, can never be harmful. The usual time for giving the morning bath is about nine or ten o'clock ; at this time digestion is not going on, as a rule, the child can be thoroughly washed, the surface brought into a glow either with the hand or a soft towel; the child can then take its bottle or breast, get its hour's sleep, and there will be still time for it to spend the best part of the day out of doors. Of course in summer, when the child should be out as much as possible, the bath can be given at an earlier or later hour to suit the circumstances. For a very young infant it is not absolutely necessary to give the child a bath in the tub; the room should be warmed to a temperature of about 75°, guarded against draughts.

The temperature of the bath, if the child is immersed, should be about 90°, but if the child is delicate and young a thorough sponging of the surface will be sufficient, and gradually it should be accustomed to the water until it will of its own accord show a liking for the bath.

The question often arises, How long after feeding should the child have a bath? Certainly not less than an hour, better if two hours should elapse after a heavy meal. Of course this refers entirely to a bath by immersion, but for a young infant that is simply sponged and nursed with breast-milk, an hour will be sufficient.

Often it is necessary to bring about a glow on the surface of the body for children who are delicate, when for some reason or another the bath cannot be given; the body can be gently rubbed with either spirits of wine or washing whiskey, to which a little salt can be added to make it more stimulating; or, if the child is very delicate, cod-liver oil can be used with rubbing.

The child should not be permitted to go out immediately after its bath, nor indeed for an hour or so, if the weather be cold; but as the day's sleep is given immediately after the bath, scarcely any mother would be tempted to take her child out. Mothers ought to make it a rule never to take a child out of doors on an empty stomach; not only will a child that has taken food before going out be better able to resist cold, but also there will be less chance of it becoming infected by contagious diseases.

> "To dare the vile contagion of the night
> And tempt the rheumy and unpurged air,
> To add unto his sickness."

If a child objects very seriously to its bath, it is far better to gradually accustom it to being immersed, and this can be readily accomplished as a child grows older by teaching it to play in its tub, and gradually fill it with water; or over the tub can be thrown a light blanket, and the child gradually immersed, gently lowered into the water.

The question of when a salt bath should be used is often asked. This, of course, is a matter which, as a rule, should be left to the family physician to decide. Salt water is more stimulating than plain water; it also has the advantage of being especially valuable in cases of chronic enlargements of the glands and tonsils, a tendency to scrofula. Children who lack muscular strength, have loss of appetite, sleep badly, are especially benefited by salt-baths. It is not necessary to obtain what is known as sea-salt, though this is usually sold for that purpose.

As children grow older the question arises as to the sea-shore and its advantages, especially sea-bathing. All children who are delicate, those that are scrofulous, those that are threatened with spinal curvature, have a tendency to become bandy-legged or pigeon-breasted, improve wonderfully at the sea-shore. As far as the bathing is concerned, surf-bathing or cold sea-water should not be used for children under three years; until that time the sea-water can be given in the ordinary tub, to which has been added a quantity of hot water to give it a temperature of at least 80°. Any one who has spent summers at the sea-shore has certainly seen a great deal of the cruel practice of carrying a screaming, struggling infant in the arms and plunging it into the sea-water. I cannot imagine a more barbarous proceeding. The sudden shock from the use of cold water, the fright, is enough to bring on convulsions. A child at the age of two years can have its bathing suit put on in the middle of the day, run in its bare feet in the sands, bask in the sunshine, get its feet wet in the cool sea-water, and receive very much more benefit than it would from a plunge into the ocean, even if it could be done without the struggle which usually accompanies this procedure. Even for infants of a year old, the sponging of the neck with cold water, dipping the feet in the same, followed by brisk rubbing, will prevent in many cases the taking of cold. I especially call the attention of mothers to this point; it will be a very valuable procedure, especially in our changeable winter-climate, if adopted every night before retiring, to prevent the many attacks of cold that are so annoying and prevalent.

Sea-bathing is to be interdicted for rheumatic children, for those with asthma, skin-diseases, and fevers. Kidney-disease, irritable lungs, a tendency to frequent bronchitis, come under the same category. In the case of heart-disease the stimulating atmosphere excites this organ to too rapid action and aggravates the disorder. Weak eyes are to be kept from the shore, where the air impregnated with salt and fine sand and glare keep up a constant irritation. The same applies to ear affections, but with exceptions, which, however, should be made only under the advice of a competent physician. Little consumptives do better in the interior, as the coast air is too stimulating for their weak lungs.

The action of the skin is so essential to good health,

that, except when a child is really ill and the attending physician has given his opinion that it should not be bathed or even washed, a simple sponging of the surface of its skin can never do any harm; of course the water can be made tepid, the room carefully guarded against draughts, and the child after being thoroughly dried not allowed to run out in the cold entries until the skin has entirely reacted. The mother will often say, " Doctor, my child has a cold; shall I wash it?" I may answer that when these precautions are taken, the sponging of the chest and throat, with subsequent friction, is the best thing she can do for the cold.

In regard to the use of cold water in nursery bathing, it is a great mistake to believe that a child should be sponged with cold water notwithstanding its dread of it and the shock which it gives to its nervous system. A child should be made to love its bath, to look forward to it with delight; it should have a big sponge to play with, and in a very short time, as it grows older, it will gladly sit in the tub with the water, splash around to its heart's content, and get sufficient exercise to avoid any chances of getting cold.

In using soap, great care should be taken that it is the purest kind, with no free alkali. That which is non-scented is to be preferred. After the child has been dried, in summer-time, its body should be powdered with a little starch or talc powder, which has a soothing effect upon the skin; or in winter-time its chest and back, the folds of the skin in the groin and axilla, can be greased with a little vaseline, just enough to make the skin soft and pliable, and also protect it from cold.

There is one point of caution which I think is in place here: a bath in tepid or cool water for a short time is invigorating; a prolonged soaking in warm water has precisely the opposite effect.

If the child is debilitated during hot weather by the prolonged heat, and a more stimulating bath than the ordinary cool bath is required, a teacup or two of cider vinegar can be added to the bath, with or without the addition of salt. In children who have delicate skins, the red spots or blotchy eruption which appears shortly after birth is usually due to too active use of soap and water immediately following their birth. This can be obviated by following the directions given when speaking of the washing of the new-born babe; but very frequently a child's skin becomes dry, rough to the touch, and needs constant attention to prevent eruptions, especially the much-dreaded eczema of children. Such children should not have salt-baths, except when advised by the physician; they need fresh air, plain food, and probably cod-liver oil.

CHAPTER VIII.

TEETHING.

The Milk-Teeth, and when they are usually Cut—Why a Slobbering Babe suffers less from its Teeth than another—Usual Symptoms—Why Mothers dread a Child's Second Summer—Rickets and what is meant by the Term—Rickets a Frequent Cause of Spinal Affection—How a Pregnant Woman may predispose her Child to Rickets—Fat Children are not always healthy ones—How to treat a Rickety Child—A Teething Child suffers frequently from other Causes than its Teeth—A Child should have Sufficient Sleep—Why so many Little Ones are wakeful—How Sleeplessness can be remedied—The Diet of a Teething Child.

THE first lower incisors—that is, the two lower front teeth—are usually cut when the child is from six to seven months old. For some time previous it has probably been noticed that the child has been restless and uneasy, that its sleep has either been disturbed or it has been wakeful. Before this time the child's secretions have been pretty well established. The tears, the saliva, will probably flow readily; indeed, such children usually cut their teeth with little trouble. It is doubtful whether all the troubles that are associated with teething are really due to that condition. There is no question but that the pressure upon the delicate nerve-pulp beneath a tooth, that is bound down by a thick capsule, and probably on top of that a congested gum, may give rise to serious trouble, amounting to convulsions or intestinal disturbance at times; or the irritation which is produced may inflame the gum, and thus starting as a sore mouth, the dryness of the mucous membrane extends to the stomach, and is the starting-point of a severe catarrh.

As a rule, children suffer comparatively little with their early teeth; possibly because, while these are being cut, they are still nursing in many cases, or they have not yet had a large amount of farinaceous food added to their diet. Then, also, the large back teeth and the eye-teeth cause far more pressure on the delicate nerve-pulp, more disturbance in the jaw, and therefore are followed by a greater degree of sympathetic derangement. This sympathetic derangement may show itself in excitement of the nervous system, especially at night, and be one of the most active causes of sleeplessness; also in its action upon the glands that secrete the fluids used in digestion, and it is on this account the second summer is to most mothers a dreaded time, as then a baby is usually weaned, and the slightest neglect in the preparation of its bottle will permit of the fermentation of the starchy material that faulty secretion prevents from turning into grape sugar.

Nature's plan is to keep the gum softened, by being soaked in saliva; it also uses the bowels as a sort of safety-valve to relieve the congested nervous system. A large watery movement of the bowels will cause a shrinkage in the gums, by depriving them of water, and often take the place of the lancet. Congestion may be present in the delicate nerve-pulp beneath the teeth, and give rise to annoyance, irritation, pain, and at the same time the gum above give no evidence whatever, by its appearance, of what is going on beneath. A child suffering in this way from its teeth will crave something to bite upon, but as soon as it takes the nipple of its bottle in its mouth, its finger, or even its thumb, and bites upon it, it will suddenly throw it from it, and show evidences of pain. The relief that comes from the soaking of the gums is very great; it is said that children who suck their thumbs seldom have trouble.

By the time a child reaches two and a half years, it should have cut its entire twenty teeth. They are usually cut in pairs: first the lower two incisors, then the upper two, then the outside two above, then below, next to those first cut, then skipping a space for the eye- and stomach-teeth, the others will come in turn. From the sixteenth to the twentieth month the eye- and stomach-teeth will be cut.

The cutting of the teeth by no means always follows in this order, nor indeed do we always find the first appearance of the lower incisors as early as the sixth or seventh month. Sir William Jenner has stated that if a child does not cut its first tooth within a year, it is an undoubted sign of rickets.

I think it well here to make a few remarks on the subject of rickets, that the mother may fully understand what is meant by the term. To some people no greater insult can be offered than to suggest that their children are rickety; to their mind, the word seems to imply some constitutional taint to be ashamed of. This is a great mistake. So large a proportion of children have rickets to a more or less marked extent that physicians feel the necessity of impressing upon the community the great importance of attention to the very subjects to which this book is devoted. Rickets is a condition the result of faulty nutrition; it is found among the rich as well as the poor. It is the result, in the latter, of exposure, starvation, neglect; in the former, the direct consequence of high pressure, nervous exhaustion, improper feeding,—in other words, wilful negligence or ignorance. The mother who fails to nurse her babe, and turns it over to the tender mercies of an ignorant nurse and a bottle, should not be surprised if her child suffers in consequence. No more, indeed, should one ignorant of the fact that starchy food will ferment and be productive of harm, be surprised to see her child develop spine-disease or become bowlegged.

By rickets we mean a disease of the nutrition of the body whereby its natural growth and development are arrested, the formation of bone is retarded, and the pressure exerted by the muscles and the weight of the body causes deformities which later in the disease become permanent. This want of bone-deposit delays the formation of teeth; the interference with nutrition causes wasting of the muscles, produces disorders of digestion, and also shows itself in affections of the lymphatic glands, the liver, the spleen, and the brain. Its causes are bad feeding, want of sunshine, dampness, want of cleanliness; and yet so gradual and slow is the process by which this faulty nutrition shows itself, that many children who are seemingly strong and hearty will manifest signs of rickets towards the end of their early dentition.

Rickets is in nearly all cases developed after birth, usually about the fifth or sixth month.

As regards the normal growth of the child during the first year of its life, the average growth is about eight inches; during the second year about four inches, and at this time it is supposed to have attained half of its full adult height. As regards weight, at the end of the first year it has gained about three times its weight at birth, but during the second year it has only increased this by about one-third.

I have endeavored to impress upon mothers the great importance during pregnancy of leading the sort of life that will give them good digestion,—plenty of fresh air and exercise,—and of a diet that will supply all the demands of nature. Otherwise at this time they are the ones who will suffer; nature will draw from them the material to supply bone to their infant. The nursing mother should also be most careful in her mode of life, knowing that the nutrition of two individuals depends upon herself. Should the supply fail, one or the other will suffer. Infants deprived of the materials that go to the formation of bone, lime-salts, will develop rickets though they may be fat; because excessive storage of fat is no evidence of health, but, on the contrary, is often evidence of faulty nutrition. The failure of the diamond-shaped opening on the top of a child's head to close before the end of the first year, the delay in cutting its first tooth until the expiration of this period, the tendency to enlargements of the glands, especially the tonsils, disturbances of the bowels, especially that form where bile does not seem to be properly secreted, the tendency to perspire at night about the head or neck, notwithstanding the fact that the child seems fat and well nourished, are all, particularly when found associated in the same case, evidences of rickets, and a child presenting these symptoms should at once have a change of air and diet, and be placed under medical treatment.

These symptoms may occur in a child that is nursed by a seemingly healthy mother; in fact, we often find that the healthiest-looking woman is the one whose milk is least nutritious. They are apt to occur in a child that has followed rapid successive pregnancies; almost certain to take place should the mother while pregnant attempt to nurse a child, and will very frequently be noticed exclusively in children who have been nursed too long. Too early weaning, especially if the bottle be made up of starchy food improperly cooked or given in too large quantities, will be a cause of rickets. Such children will develop, besides the other evidences above noted, large bellies distended with wind, vomiting occasionally sour matter, alternating diarrhœa and constipation, and the stools will most frequently be found either chalky or putty-like, and containing quantities of mucus. If a child is late in cutting its first teeth, though rickets may not be present, there is something faulty in its condition. By late, I mean if it has not cut its front tooth by the tenth month. Such a child should be carefully watched, its diet freely supplied with bone-forming material; it should be given Mellin's or Horlick's food with its bottle, or (Trommer's) extract of malt, without hops (a teaspoonful to the bottle). It should have a salt-bath daily, its muscles should be exercised by daily rubbing or massage, and under no circumstances should such a child be allowed to stand on its feet until it has demonstrated its ability to do so after the gradual strengthening of its bones and muscles by creeping.

A great mistake is often made by permitting a child to creep in one position. These children also should be watched with care, that they receive no sudden shock, or blow, or fall. Spinal curvature may be the result; distortion of their hip-bone or pelvis,—a serious matter if they are girls,—or hip-disease may develop most insidiously. Diseases of the lungs should be guarded against, as bronchitis, pneumonia, etc., as the difficulty in breathing will press the weak ribs out of position and keep them so, making them pigeon-breasted. It is said that those children who cannot digest starchy food are more apt to suffer from rheumatism.

I have dwelt at length upon this matter in order to

show that what is usually attributed to difficult dentition —the head-sweating, digestive disturbances, sleeplessness—in many cases may have nothing whatever to do with dentition, the difficult dentition being simply a symptom in the course of the disease. Malnutrition, rickets, is the true cause. There are certain disturbances, mechanical ones, that are caused by the pressure of the teeth upon the nerve-pulps beneath, that are especially noticeable in children of highly nervous organization, those whose parents are brain-workers, and in these cases we are apt to have disorders of digestion, malnutrition, brain-excitement, resulting in sleeplessness and possibly convulsions. Especially do these conditions manifest themselves during the cutting of the back teeth, as the molars, the eye-teeth, and stomach-teeth; the eye-teeth, so called because their cutting is usually accompanied by more or less disturbance in the circulation of the brain and the upper parts of the face, giving rise to excitement, to sleeplessness, or to catarrh in the nose or in the eyes, and the stomach-teeth because their cutting is usually accompanied by disorders of digestion.

Sleeplessness and irritability seem a constant accompaniment of the teething process. A new-born infant sleeps from fifteen to eighteen hours out of the twenty-four; a child of two years should sleep with little interruption at least ten or eleven hours at night, and from one to two hours in the middle of the day, but sleeplessness is not always an evidence that the child is suffering. Habit has much to do with this condition. A child should be prepared for sleep; its hours should be as regular as clockwork, and under no circumstances should it be allowed to pass its sleep-time. Especially is this rule important during the time of teething. The reasons that many children are sleepless are, first of all, our city children are of an excitable temperament; they do not get sufficient fresh air or exercise, and in consequence the fatigue which should naturally invite profound and peaceful sleep is accompanied by a degree of excitement that prevents this.

Sleeplessness from this cause can be remedied by attention to these matters. It is produced by exciting a child just before its sleep-time, especially after it has taken its bottle; its bottle or bowl of food should be given the very last thing. The drowsiness which follows this meal, if once disturbed, will cause a wakeful night. Among the poor, the father comes home tired from overwork, and the family retire together at an early hour; with the well-to-do, the lights are all lighted, the older children have their romp, or the baby, if it be the first, has to have its half-hour of play with the father, and possibly the excitement of its brain may cause a restless first sleep and wakefulness for the greater part of the night. For such cases a hot foot-bath will have a most soothing effect, or, if they still resist, the morning bath can be postponed until evening, giving a hot bath at this time, and a sponging in the morning instead. In these cases, a bottle of food as warm as the child can take it, consisting of Mellin's food and water, the proportions being a tablespoonful of the food and water to fill the bottle, which will put the child to sleep. If the wakefulness is due simply to excitement, the child seeming perfectly well otherwise during the daytime, medicine should be avoided, unless specially ordered by a physician. The habit of giving bromide, for no other reason than simply wakefulness, is a mistake.

Another cause of sleeplessness is insufficient nourishment; we sometimes see such cases, where the mother is gradually losing her milk : its bulk remains probably the same, but it is losing its richness or quality. Children fed on condensed milk alone are sometimes underfed, from the fact that the milk is given too highly diluted. One part to twelve of water should be the strength up to the second or third month; after that, about one part to ten or eight, as the child grows older, or, better still, increasing the quantity of cream. But I fear that among those who will be readers of this book the greater cause will come from overfeeding. A restless infant, and especially a child about a year old, who tosses in its sleep, cries out, mutters, dreams, is one in all probability who is receiving food in excess, or is certainly not getting rid of the surplus in a proper way; if the mother notices that a sleepless, restless child, at whatever age it be, has a furred tongue, that its breath is heavy, that its urine is scanty, high colored, or that it passes water very frequently, probably wets its bed at night, she will know that the child needs a laxative, to be followed by a change in its diet, —a dose of Husband's magnesia, about a quarter of a teaspoonful, or, probably better than all, a dose of castor oil, followed next day by a limited amount of milk, substituting chicken-broth and avoiding oatmeal or beef-juice until the bowels become more free. This form of sleeplessness is most apt to be noted at about the end of the first year. With some children the constant diet of oatmeal without any variation, the daily use of the expressed juice of beef, and at the same time large quantities of milk, which at this period is not often much diluted, will bring about the form of digestive disturbance just mentioned, and a degree of nervous excitability that is often referred to the teething process alone.

We often hear of children who are said to be suffering when cutting their back teeth, and upon investigation find that they are taking about a quart of milk a day, with a tablespoonful of Mellin's food in each bottle, an ounce or two of the expressed beef, possibly a bowl of oatmeal food, and in addition are constipated, get no fresh air,—in fact, are city children, surrounded by all the disadvantages that our changeable and treacherous climate affords; viz., overheated houses and impure air. A child of this sort, if given chicken-broth instead of beef-juice, less Mellin's food, thin bread and milk instead of oatmeal, allowed to drink freely of water, and given an occasional laxative, will soon cease to suffer with its teeth.

Sleeplessness may arise from pain; especially is this the case in bottle-fed children, who suffer from neuralgia, muscular soreness, possibly vague rheumatic pains, supposed by many to be caused by the fermentation of the sugar or starches which they are unable to digest. A child that is fed upon too much sugar will develop acidity; the same with starch; this leads possibly to rheumatism, or rheumatic neuralgia, called growing pain, and eventually to rickets, restlessness, sleeplessness, or sudden starting, soreness to the touch, no desire for exercise, peevishness, or possibly extreme pain upon movement. When the child is lifted suddenly, or is handled while dressing, it will cry,—all important symptoms of this condition.

An extra-sensitive nervous system will probably cause neuralgia through the jaw and head, earache, in some children, from tooth-pressure; this cause is recognized from the fact that these children will avoid anything placed in the mouth, even the nipple of their nursing-bottle. The gums may be slightly swollen, but not inflamed or in themselves tender, but the least pressure upon them with the finger will cause intense suffering. In a case of this kind, if severe,—the child restless, starts in its sleep, refuses food,—the gum should be lanced. A cross-cut that will open the capsule of the tooth will in a moment relieve the pressure. If this is impossible, a hot bath or hot foot-bath should be given, also an enema of warm water, and if the child still suffer, and the mother is away from the doctor, she may give two grains of bromide of potassium, or five drops of the elixir of the valerianate of ammonia, the latter to be repeated in the course of an hour if necessary. These are the cases in which the bromide of potassium, if given judiciously, is of the greatest value; they present the purely nervous type, and the nervous system should be soothed by precisely the same mode of treatment that one would adopt for a severe neuralgic headache. The bromide may be given with a teaspoonful of syrup of lactucarium, or a teaspoonful of orange-flower water. If the excitement is great, the child of a year old should take about four grains of bromide; should a convulsion threaten or occur, the proper treatment is warmth and counter-irritation to the whole body, a hot mustard-bath as hot as can be borne by the mother's arm, or hot mustard-poultices to the legs, abdomen, and back, which some prefer; an enema of two ounces of water, about the temperature of 100°, containing one teaspoonful of the tincture of assafœtida. After the bath the child should be thoroughly wrapped in a blanket, and the only food given it for a time should consist of the blandest kind, such as barley-water, wine-whey, or chicken-tea. Mothers should not give their children ipecac in this condition, unless there is decided evidence of indigestion, it being very irritating, and the strain produced by vomiting may be productive of harm. Ten drops of the aromatic spirits of ammonia, or ten drops of brandy, whiskey, or gin, in two tablespoonfuls of warm water, are far better, when given to the child as soon as it can swallow; if there is any offending matter in the stomach, it will act sometimes as an emetic.

There is another cause of restlessness and sleeplessness from teething, that due to the congestion of gum and inflammation of the mucous membrane of the mouth. We more frequently find this condition when the back teeth are being cut, and in children who secrete but little saliva. The gum is found swollen, red; the mouth is dry; the child will turn away from more solid food, and will eagerly drink water to allay the irritation in its mouth. Frequently, if it is during the second summer, it will be accompanied by diarrhœa, from the fact that the mucous membrane extends to the stomach and bowel. Very often small ulcers form in the fold of the cheek or the surface of the gum, become coated with secretion, and are exceedingly painful. If there is disturbance in digestion, little round vesicles, which burst and form ulcers with a grayish coating, will form on the inside of the cheek or the lips, and on, or beneath, the tongue. These aphthæ will annoy the child extremely, and produce sleeplessness; they are the result of indigestion, not of teething, though frequently associated with it. The physician will probably give a little calomel and soda to correct the digestive disturbance, or the mother can give some magnesia, or from ten to twenty drops of spiced rhubarb with a teaspoonful or two of soda mint, given once a day until the bowels become regulated; at the same time the child can have its mouth washed gently with a soft piece of linen, with a solution composed of a pinch of borax, a teaspoonful of glycerin, and a tablespoonful of rose-water.

The congestion of the gum of which we have been speaking causes the child to crave salty food; indeed, this is an effort of nature to relieve this condition by inducing a flow of saliva. Lancing the gums will often be of great service, by the relief it gives to congestion; instead of this the child should be given something to bite upon, and it will probably obtain relief in that way. The small bone of a well-boiled ham is salty, and the child will eagerly suck it, and frequently bite it. A chicken-bone, slightly salted, is most useful, or it can be allowed to bite on a piece of rare roast-beef, and possibly the eagerness with which it will seize upon it may cut the gum from below the sharp points of its tooth. Children who suffer thus from their gums often get relief from Mellin's food, tied in a rag and given the same as the ham or chicken-bone.

To sum up, the object to be secured is the free flow of saliva; it will relieve congestion and soften the gum. It is a mistake to rub a gum under such circumstances by hard dry friction, but if the little one suffers so as to disturb its sleep, the mother's finger dipped in the syrup of lactucarium, or even in paregoric and glycerin, can be gently carried over the tender and inflamed gum, and, by gentle pressure, soothe instead of irritate; and now and then a little firmer pressure may allow the point of a tooth to force its way through.

In these cases, a hot foot-bath again has its advantages, by relieving the congestion of the head and mouth; or, if the child is constipated, the operation of a laxative will be followed by relief to its congested gum. The latter will shrink, and a point of the tooth will emerge from its captivity.

CHAPTER IX.

DIET AFTER EARLY DENTITION.

The Necessity of a Careful Selection of Diet—The Advantages and the Disadvantages of taking a Child to the Table—The Necessity of discriminating between Children as to their Diet—Why what is Meat to one Child is Poison to Another—Exercise is Essential.

AFTER a child has cut its twelve teeth, it is well to give it more solid food than what it has been accustomed to take. Probably the milk diet has been continued, or the bread and milk, up to this period, in addition to the mutton, chicken-broth, or beef-soup, and our endeavor now should be to encourage digestion of more solid food

DIET AFTER EARLY DENTITION.

by adding it more gradually to the diet to which the child has been accustomed. In almost every house is kept on hand what is known as *stock;* this forms a very valuable addition to a child's dietary, as it is nutritious, palatable, and can be flavored so as to make a change. A child about two years old can have a more solid midday meal, composed of a rare piece of tenderloin or piece of juicy mutton-chop, with some well-boiled rice or a thoroughly roasted dry mealy potato, in addition to its soup. For its breakfast, instead of the bottle, a child of this age could take a small plate of about a tablespoonful of cracked wheat mush, which must be thoroughly boiled or steamed, or oatmeal, or yellow corn-meal, or white grits of moderate consistency, but thoroughly boiled, and milk. If a little salt is added to these while boiling, flavored in this way, it is preferable to giving sugar. I think, indeed, this is a better plan than mixing with sugar or the other alternative of using salt and butter; butter in that way will certainly upset a child's digestion.

A soft-boiled egg, with bread and butter and a tumbler of milk, is about the best breakfast a child can have at this age. If it sits at the table and is taught to eat slowly, it will not become dyspeptic. Between times, if a child is thirsty, a drink of milk is admirable. With its dinner, water is probably better; I certainly have found it so in children who are of a bilious habit. It is a great mistake to give a child sweet things before its meal; after dinner it might be allowed to have some mild dessert, custard or pudding that is light, sponge-cake, or a baked apple, or, indeed (after it has cut its twelve teeth), a piece of ripe, raw apple or peach will have a good effect on its bowels. The great advantage of bringing a child to the table to eat with its parents is that it is taught to eat properly, to masticate its food thoroughly, which is the great secret for avoiding dyspepsia; and also the child can be so trained, if its parents wish, to see things that it knows it should not have by asking for them. It is a great mistake to spoil a child so that it will refuse when at the table the proper sort of food, and cry constantly for the impossibility. A taste of this or that at the table may not result at the time being in bringing on troubles with digestion, but the parents are sure to suffer for it in the future. We so often make the mistake of believing that children are rendered strong and hardy by inattention to these matters of diet and clothing, that carelessness makes the child hardy, that the child that is strictly brought up is usually a sickly one. I acknowledge that at times great mistakes have been made by over-carefulness, that the scientifically brought up child is not always the most healthy. We can overdo these, as we can everything else. It is always well to make your list for the child's bill of fare as large as possible, and give it its choice; but always adhere to the lines of digestibility, and avoid those articles that every one knows are absolutely indigestible or at least harmful and irritating, such as veal, pastry, unripe fruit, sweets, and do not believe that because children sometimes escape the dangers that indulgent parents bring upon them, the words of advice only come from those who are too highly scientific in their ideas!

The highly-educated classes and those who live by their brains alone are notoriously dyspeptic, and the children of these undoubtedly inherit the weak digestions of their parents as they do their more highly-organized brains and possibly larger heads. They are more subject to acute brain-troubles from this cause, and they are certainly more subject to all the influences which produce intestinal disorders, and cannot possibly digest the same food that will agree with a child of the same age, of the less intellectual and the laboring classes. I think this is a matter which every mother should thoroughly understand; it explains to her why the one child will thrive on food that would be poison to another; it will prove to her beforehand, without the necessity of an experiment, that her child would not be made more hearty by feeding on the boiled potatoes, soggy bread, corn-starch, or apple-dumpling that has failed to kill her washerwoman's child. Nature has so provided the food that it is not all concentrated nourishment; in grain, in fruit, in meat, the easier digested portions have with them certain materials that are harder to digest; in a mixed diet the various substances have different degrees of digestibility, and in this way the digestive juices come more closely in contact with the food, owing to its bulk, and the muscular contraction of the stomach and intestines is promoted by the mass of material which passes through them. It would be, then, a mistake to feed a person in health entirely on concentrated nourishment; the intestinal digestive juices would fail to be secreted, the liver would become congested and diseased, the bowels would be coated with mucus and would fail to act, and the refuse from the destruction of tissue which is carried off by them would remain behind and poison the system. But during illness, when digestion is checked and when it is necessary to supply in small amounts the most concentrated nourishment to sustain the system of the patient, then the well-known preparations of peptonized foods, of concentrated meat-extracts, are not only most valuable but often absolutely indispensable.

After it has cut its twelve teeth, the child will naturally desire a greater variety of foods, and within bounds these desires may be satisfied. Its bill of fare will include articles that are easy of digestion,—mutton, beef, sweetbread, fowl of various kinds, soft-boiled eggs, eggs scrambled, or light omelet; vegetables, as cauliflower, young beans, beets, potatoes,—the latter well boiled and mashed, or roasted,—asparagus; by all means avoid parsnips, cabbage, turnips, celery, etc.; for fruits, baked apples, raw apples scraped, oranges—of course only the juice, grapes—avoiding the skins and seeds, peaches—provided they are not picked green and then ripened,—and avoid berries of every kind, also pineapples, green apples, etc. Figs, dates, and raisins are all bad for children at this tender age.

As thorough digestion and nutrition require fresh air, mothers should bear in mind the importance of letting those children that eat heartily keep out of doors as much as possible. The over-feeding, even with nutritious food, in our hot American houses during wintertime, is undoubtedly one of the great causes of disturbances in the liver which are so frequently met with. It is the same process almost that is used in the making of the *pâté-de-fois-gras*. The Strasburg geese are over-fed and over-heated. Constipation, which is so common in these bilious children, cannot be overcome by laxatives or purgatives, and mothers should avoid their use or their abuse. Let a bilious child be allowed to drink water freely, to have a good run in the open air with its clothes sensibly loose, ride on its tricycle, play with its hoop, jump rope,—the biliousness will rapidly diminish and the bowels become regular.

CHAPTER X.

ON THE BOWELS

What constitutes a Normal Condition of the Bowels in a Child—More Grease than Powder should be used for Young Children—What causes Disturbances of the Bowels—Different Causes of Constipation—How Constipation can be overcome—Of what a Child's Diet should consist, and how it can be regulated when it is suffering from Bowel-Complaints—A few Useful Remedies—Exercise a Specific for many Forms of Constipation.

It is a very important matter for a mother to know what constitutes the normal condition of the bowels of her children. At birth, the intestines contain an accumulation which is carried off by the first milk, this milk being laxative in character. The passages will then assume a yellow color, and in about two or three weeks will begin to get gradually solid, in part. The passages should have little odor, should not be sticky, nor contain mucus, and should be liquid. Up to three or four months there should be two or three or even from four to five movements in the twenty-four hours; though many children are perfectly healthy who have but one movement in the twenty-four hours even at this tender age. As soon as a child has had a movement from its bowels it ought to be changed, and it is well to thoroughly anoint the parts with vaseline; anything that causes irritation of the skin—undue acidity of the passages, or too frequent passages from the bowels—will rapidly cause soreness, unless washed and then thoroughly greased. It is a very great mistake for mothers to use rubber or oilcloth with the diapers, as these will draw or keep the parts in constant perspiration. It also must be closely watched that they are not washed with soda. Grease is far better to use on children's skin than powder. A little white vaseline, as before recommended, tallow, simple cerate, or lanoline, will make the best ointment.

As the child grows older the movements become less frequent; by the time that it is six months old it will have about two passages in a day. These passages, of course, are not always exactly the same for a nursing child; indiscretion on the part of the mother, the use of purgatives, of certain articles of diet, will so affect her milk as either to cause vomiting or purging in her infant. The child's movements may vary in color, becoming greenish or variegated, or they may become dark or thin, watery, or again they may be very much constipated, hard, round, surrounded by mucus, or again the passages may be white, looking like pure cheese.

Constipation may be evident from the very earliest movements of a child's life, and be extremely difficult to overcome.

We will study the two conditions,—constipation and diarrhœa,—and endeavor to point out the cause and remedy in each case. Owing to the smallness of the bowel and to its position in the infant, there is not that accumulation which takes place in the adult, in the lower bowel, which acts as a reflex, or sympathetic, irritant and gives occasion for the muscle to contract to have a movement. This reflex irritation, or desire to have a movement of the bowel, has a marked tendency to return at certain hours,—that is to say, to become a habit,—and when properly regulated at that period of a child's existence when the habit can be established,— when the child is a year old, for instance,—the tendency which might have existed previously to constipation will rapidly disappear, especially as from this time forward the child's bowel gradually changes its condition and allows a greater distention to take place. Constipation in the infant at the breast is, indeed, one of the most difficult problems to solve that I know of; at times it seems to be a hereditary condition, a lethargy of the intestinal nervous system, which fails to respond to irritation. The contents of the bowel are composed of many materials which have escaped digestion, curds, mucus, the secretions which have been thrown into the bowel to help digestion, or to lubricate it; the secretions from the liver especially, which are intended first to help digestion by aiding the absorption of fats, second to prevent decomposition, as bile is a great anti-putrefacient, and third, the bile contains the refuse, the ashes, that have been thrown off by the liver from the use and destruction of tissue. These materials, if not thrown off by the liver, are poisonous, and give rise to the symptom known as biliousness. Now, mothers must not make the mistake which is so prevalent at the present day, to imagine that constipation always means faulty action of the liver. If the child is constipated, the passages white or chalky, much flatulence or colic, and the tongue is coated, and the urine stains the diaper with a reddish hue, then there is a decided want of action of the liver, and the constipation has probably a cause which can be removed by appropriate treatment; but frequently the liver itself may be working perfectly well, while a catarrh of the bowel, a result of cold or indigestion, may be seated at that part where the bile flows from the liver into the intestine, and the flow of bile be checked.

The symptoms will be very much the same in these two cases, although it is obvious that the cause is different, and the mother who recognizes the condition but fails to grasp the difference and doses her child without the advice of her physician, may do an immense amount of harm.

It is on this account that throughout this little work I have insisted upon submitting all such matters to the family physician. I am opposed to household works on medical treatment, except as far as they give general information and are part of a liberal education. Every physician knows that with an educated mother, who is thoroughly in accord with the doctor in his endeavors to understand the child's condition and treat it properly, the chances of success are greater than when the parent is ignorant of those many details which works of this sort impart.

Constipation may be due to the character of the food, to the want of secretion in the intestine, or to the failure of the intestines to contract and propel the material towards their outlet,—what is known, technically, as peristaltic action. An infant at the breast, or bottle especially, may be constipated and every endeavor fail to give it regularity, yet as soon as it begins to take solid food the bowels will immediately become normal. A mother should recognize these different conditions in order to be able to counteract them. If the constipation depends on the character of the food, its indigestibility, or the rapidity with which the water is absorbed, leaving an excess of solid or curds behind, the movement of the bowels will be of cheesy character, putty-like, the masses hard, lumpy, possibly not differing very much in color from what they ought to be, but they are surrounded by mucus,—in fact, very much like putty. The movements may be infrequent, or there may be quite a number of very small movements, showing an irritability of the rectum, and this may be accompanied by what is known as the diarrhœa of constipation. This form of bowel-trouble is found especially in children who are weaned from the breast and who are on bottle-food, and its treatment is to give freely some of the broths, such as mutton, chicken, or beef, a cupful or two during the day; to avoid a heavy meat diet, or one composed exclusively of milk; to give the child its bowl or bottle of thin boiled bread and milk, using stale baker's bread and straining, and making it thin enough to pass through the nipple of the bottle, which should be made larger than that ordinarily used. The child should have some preparation of malt, or pepsin, to aid its digestion. A teaspoonful of wine of pepsin in a claret glass of water immediately after eating or just before, if the appetite is at all failing; and if still the masses of matter passed show indigestion, the child should have about ten drops of the aromatic syrup of rhubarb with five drops of the wine of ipecac, every night, in a claret glass of water, until the passages show that by their healthier color the bile is being secreted. If the digestion seems still to be weak, about half a teaspoonful of the solution of lacto-peptine can be given after food.

For those children who have been weaned from the bottle and are taking thicker foods, that is, solid diet, it is a mistake to give water too freely with their meals. The water should be given between meals, as in that way it not only gives water to the system, which needs it, but it washes out the stomach and bowel of undigested matter and mucus, aids, by its mechanical action, the passage of materials which should be discharged, and relieves constipation. It is on this account that a glass of water the first thing upon rising in the morning is recognized as a laxative. If the child has much straining, it will be noted that the passages are streaked with red blood; this is caused by congestion of the mucous membrane, and can be avoided by the use of some soothing enema. A small hard-rubber syringe holding about one ounce can be filled with the following:

   Sweet oil, a tablespoonful ;
   Warm water, 100°, "
   Pinch of salt.
    Mix thoroughly.

This, given to the child at the time when its bowels should be opened, will give something for the muscle to contract upon, and clear the mucus out of the lower bowel. If the mucus and streaks of blood still remain, a thin starch-water, boiled, should be used in the same way.

Another form of constipation is that which is simply due to want of propelling power in the bowel. In these cases a child will go for several days without a movement, although apparently in good health; when moved, the passages are to all appearances perfectly normal, large, well formed. It is astonishing how much can accumulate in the bowel. Indeed, it is probable that most of the over-fed, fat children of the well-to-do in our large cities, who take little exercise, and that consisting of a daily parade with the nurse, so dressed that it is impossible they can take the exercise required, have their bowels much distended with matters that finally undergo decomposition and are the causes of the blood-poisoning which foul breath, furred tongue, loss of appetite, languor, drowsiness, indicate, all grouped under the synonyme of biliousness. These flabby children are constipated simply through want of propelling power in the bowel or in the muscles of the abdomen. For these the treatment is generally divided as follows: the external, the dietetic, and the use of medicines. The external treatment consists of manipulating or kneading the abdominal wall. As this condition exists in individuals from infants at the breast, we might say, until old age,—for it is almost as common with the parent as with the children,—the treatment more or less modified is useful for all. The child, after its bath, should be placed on a blanket, on its back, on the mother's lap, and the abdomen gently rubbed, beginning by placing the palm of the hand upon the navel and rubbing with a circular motion gently but firmly until the surface is quite in a glow, each movement increasing the size of the circle, like the rings in a pond after a stone has been thrown into the water. Soap liniment can be used, or cod-liver oil, or sweet oil, or even castor oil, externally, if constipation be marked. As the contents of the bowel descend on the left side, the movement should be from left to right. One good rubbing a day will frequently be followed by a movement of the bowels. This may be still further increased by placing the hand in the same way and shaking the abdomen. Of course, exercise will be of great service in this form of constipation. For an infant this can be secured by gentle manipulation; for an older child, passive exercise and the encouragement of out-door sport.

The next form of constipation will include that which is due to derangements of digestion, probably produced by too highly stimulating food, or food which has a tendency to ferment and produce gases. Mucus will be secreted in the bowel when the intestine is irritated by food or cold, and instead of producing diarrhœa, which would have been the result had the intestines been inflamed, may cause attacks of vomiting, colic from flatulence, the expulsion of gases which have an extremely offensive odor. The child's abdomen is swollen with wind, and the passages usually are offensive, possibly fluid, or they may be hard, dark-colored, and infrequent, accompanied also by a coating of mucus. Children that are fed largely upon eggs frequently suffer in this way, or those who receive a large amount of starches not sufficiently boiled, which remain undigested in the bowel. The children lose their appetites, become peevish, restless, suffer with inordinate distention of the abdomen, and, finally, if the cause is allowed to continue, get catarrh of the stomach and intestine, and obstinate diarrhœa will follow. Of course, the treatment in cases of this kind is obvious; the diet should be regulated, and such harmless laxatives used as will relieve the child of the offending elements and reestablish its digestion. Stop all solid food for a day or two, also milk and starches; put the child on broths, and give it a dose of castor oil, a teaspoonful or two. A tablespoonful of liquid soda-mint can be made quite warm, castor oil mixed with this, when it floats to the surface it can be readily skimmed off and given to the child by the spoon. Sweet oil can be used in this way for an infant, instead of castor oil, if preferred. If there is much vomiting in these cases and food is not tolerated, such as broths or barley-water, the nourishment can be given in small quantities, using Valentine's or other beef extract. Five or ten drops of whiskey or gin in warm water can be given every fifteen or twenty minutes, or two or three drops of aromatic spirits of ammonia, or gum arabic water and lime-water can be used for a time until the vomiting is relieved. Dr. Walker, of Brooklyn, recommends the following:

Creasote, two drops;
Glycerin, two teaspoonfuls;
Water, a small tumblerful.
A teaspoonful of this every hour.

And if there is much colic, he advises—

No. 1.
Aromatic catnip-tea, two tablespoonfuls;
Tinct. asafœtida, ten drops;
Syrup (simple), two tablespoonfuls.

Or No. 2.
Aromatic spts. of ammonia, fifteen drops;
Essence of peppermint, ten drops;
Glycerin, a dessertspoonful;
Aniseed-water, two tablespoonfuls.

A quarter to half a teaspoonful of one of these in water every fifteen to thirty minutes until relieved, if necessary.

Use hot foot-baths with mustard and water, and apply mustard-poultices containing half mustard and half meal to the abdomen, or if the pain should still continue an enema of warm water, about half a pint or less, or one of hop-tea. When colic and constipation exist, a mother should never give a purgative without consulting her physician, as twisting or constriction of the bowel may be the cause, but recourse can be had to the above-mentioned treatment, which can never be harmful. In speaking of constipation, I may also mention the fact that when it is obviously due to a want of expulsive power in the child the use of suppositories is often of value, and when cautiously given they can be used for some months until the child has gotten into the habit of having the bowels moved daily at the same hour. These suppositories can be made either of Castile soap, or coca butter; or the gluten suppository made by the Health Food Company is often very useful. It is a mistake to give a child laxatives and purgatives as a routine practice. Vary the diet; change its bottle-food from one thing to another; encourage it to play; give it out-door exercise. These will often in themselves be sufficient. For older children *bran*, as crackers, is often laxative, or it can be made into bread for a change.

PART III.

CHILDHOOD.

It is proposed in this section of our little work to give to mothers and nurses a plain statement of the causes and method of nursing of the more important disorders and diseases of children. By disorders we mean simply functional disturbances; by diseases we mean those disturbances that are accompanied by some structural changes.

It is not necessary to dwell at length upon the appearance of a sick child. We have endeavored to form a picture of a child in the enjoyment of perfect health, —an infant with all its functions working in perfect accord, whose sleep is soft and gentle, who awakes bright and cheerful, who eats with an evident relish for food, and who becomes drowsy as digestion begins; one whose eyes are clear and bright, its skin soft, its flesh firm, bears evidences of health. But when the child becomes peevish, restless, or drooping, uneasy after eating, starting in its sleep, when the eyes lose their brilliancy and become encircled with dark rings, its skin hot and dry, and possibly the hands and feet cold, its flesh flabby and soft, and the rotundity of its form marked by a tendency to angularity, showing a loss of the cushions of fat, it does not require much experience to recognize such as a sick child. Whether or not this deviation from health is simply functional, or due to disease, is not the question we have to deal with; that rests with the doctor; but as he depends on the mother for a true recital of those symptoms upon which he bases his conclusion, a habit of accuracy in observation, one of thoroughness in investigation, should be cultivated by her. This, together with the thorough carrying out of all the details of treatment, not as a mere machine but as an intelligent being, one who is capable of exercising judgment, form the essentials of a good nurse, and this every mother should seek to be.

CHAPTER XI.

ACUTE AND CHRONIC NASAL CATARRH.

Affections of the Mucous Membrane of the Nose, Acute and Chronic, in Infants and Children — Their Prevention and Treatment.

FOR the answers to the following questions, which are interesting to mothers in this connection, I am indebted to Dr. Alexander W. MacCoy, of this city.

How to treat an ordinary cold in the head, with household remedies for a child over six months of age ?

A cold in the head should never be neglected. At the beginning, an attack (ushered in by fits of sneezing and slight feverishness) can often be arrested before the watery flow begins by the prompt use of quinine suppositories, from one-half to two grains each, according to age, introduced into the bowel once or twice in the day; also small quantities of sweet spirits of nitre in iced water (or for older children in lemonade), taken freely. Sometimes large quantities of cold water taken will act so promptly on kidneys and skin as to quickly relieve the nose. For some delicate children whiskey and water in proper doses may be used. Hot (mustard) foot-baths upon retiring is a time-honored and very efficacious treatment if the extremities are well protected during the night. If the nostrils show watery and mucous discharge, the nasal chambers should be looked after, and the stuffiness and stoppage to breathing through them must be combated by lubricating them within with bland oils dropped in or snuffed up, or, better, used as a spray in an atomizer or vaporizer. Fluid cosmoline is one of the most agreeable and effective substances available; combined with a one-per-cent. to two-per-cent. solution of cocaine it is most efficient in relieving the nasal distress and stoppage in the nostrils. Plain cosmoline, warmed, and applied in and around the orifices of the nostrils, will greatly add to the comfort and repose of infants. Very often infants suffer more from the accumulation and adherence of the secretion, which soon dries up and renders the small nasal orifices stiff, uncomfortable, and occluded. The frequent application of a weak solution of baking soda (bicarbonate of soda) to the nostrils on a soft rag or absorbent cotton will easily remove this dried secretion, and if immediately followed by the free use of oil or cosmoline, will prevent this annoyance from recurring. Guarding the child from over-heated rooms during the day, and especially at night, with the judicious use of quinine suppositories and the application of some oil in the nose, will generally relieve and cure an ordinary cold in the head in one or two days.

1. *At what age do children show symptoms of chronic post-nasal catarrh? What are its earliest symptoms?*

The time at which a discharge from the head occurs, either from the nose or throat, may be coincident with the child's birth. An acute cold in the head often develops during the first week of life, and, followed by a succession of other attacks during the earlier months of infancy, may give rise to symptoms of chronic nasal catarrh as early, at least, as the first year. (Depending upon some hereditary taint, the symptoms are familiarly known as "snuffles," and date from birth.) A post-nasal catarrh—a dropping of mucus or phlegm from the head into the throat—is only a symptom of a nasal catarrh; without a nasal catarrh existing at the same time there can be no post-nasal discharge. It is probably true that, in young children, for a long time there is no actual inflammation behind the palate,—post-nasal space,—in discharges from the head into the throat, but the secretion comes from the back of the nose and slides down the palate into the throat; the throat may look red, but this is caused solely by the mucus lying there or constantly passing over it. An acute cold in the head will give rise to a discharge into the throat, which in children will be much more noticeable in the recumbent position. This is also true in the chronic form. One of the earliest symptoms manifested in an infant or child suffering from nasal catarrh is a short, irritative cough, generally very persistent and occurring chiefly at night, disturbing the child's rest to a distressing degree. A cough arising from mucus flowing out of the back of the nose and down the throat, finally tickling the " speaking-box" (larynx), is not amenable to ordinary treatment by cough-syrups, etc., but requires the removal of the secretion from nose and throat. The cough comes on generally after the child has been asleep for some time and the recumbent position has started the secretion downwards in the direction of gravity. During the day, the usual watery or mucous discharge from the nose is seen, and frequent attempts at swallowing may be noticed. It is quite unusual for parents or nurses to give much consideration to a cold in the head, and it is more rarely understood that this disturbing cough at night *is dependent upon the state of the nasal passages.* If a cold in the head is overlooked or neglected, as is so often the case, the constant discharge of mucus into this space behind the palate and its continuous flow downwards will, in time, produce an enlargement of a gland at the top of this space behind the palate, as well as cause a pharyngitis, which gives rise to a serious condition and makes what is often called "a weak throat." And the trouble does not end here, but often causes change in the voice, rendering it husky and hoarse, and, if left to itself, in many cases causes inflammation of the windpipe and also of the bronchial tubes; and in very many subjects too often renders them weak-chested and liable to acquire some grave pulmonary disease. In the space behind the palate, up in the roof, where this gland of which I have just spoken lies, an enlargement of the gland develops; in some cases where this occurs the tissue hangs down over the nasal openings behind, and causes impeded breathing through the nose, producing mouth-breathing, and causing the voice to become flat and nasal in character. This nasal voice is also dependent upon obstruction in the nasal passages themselves, and gives the true explanation why we are said, as a nation, to " talk through our noses."

2. *Are there any precautions that can be taken to prevent it?*

The precautions necessary to be taken to prevent a chronic nasal catarrh are comprised under the hygiene of infants and children, and the adaptation of children, from birth, to their environments. It is a lamentable fact that infants and children suffer the greatest neglect of proper care of the nasal passages. From the first bath after birth onwards, the mucous membrane of a child is put on the defensive. Too frequent bathing (daily) of infants and children, with much too warm water, in overheated rooms, followed by too little friction of the body, is a fruitful cause of "colds in the head." The important task of bathing is generally given over to the " child's nurse" after the first few months. This position is frequently filled by a young girl remarkable chiefly for her inexperience and stupidity. The little innocents are at the tender mercies of such persons, not only in the matter of bathing, but in that of dressing, undressing, proper regulation of amount of bed-clothing, and ventilation of the bed-chambers. If the natural guardians of children would give more attention to the details of these daily matters of so great interest to the physical welfare of their offspring, the prevalent nasal catarrh would become much less frequently seen. Overheated bedrooms (furnace-heat night and day) and too many bedclothes contribute greatly and promptly to an attack of cold in the head, or add to one already present. The child, after being asleep for one or two hours, is found bathed in sweat; such discomfort renders it restless, and it naturally seeks relief, and is soon outside of the coverings. This condition of affairs, kept up night after night, soon renders the skin relaxed, and greatly enhances the risk of taking cold from the rapid evaporation from the body. The selection of suitable bed-clothing is an item of great importance to the child's welfare. Light, porous blankets are the only bed-coverings advised. Luxurious eider-down and wool comfortables, and all coverings of impervious character, are to be avoided. Eider-down is especially liable to be used, because of its light weight, when the atmospheric conditions do not warrant its use. It is only suitable for Arctic climates and bedrooms without heat. A great amount of exercise in the open air (life in the country) is one of the best preventives of nasal catarrh.

3. *Should a child be taught to blow its nose? and should water be snuffed up by it to aid it, or its nose washed out with a spray daily?*

A child should be taught to blow its nose if done properly, but as the nose-blowing is generally done it is conducive of harm to the middle ear, and is thought by some to greatly increase the chances of earache and running ear. But if the handkerchief is simply placed under the nose and the discharge blown into it, there can only good results follow, viz.: emptying of the respiratory tract of the nose, and the promotion of free nasal respiration,—a very important function. Water should never be snuffed up the nose by children (or adults), except in case of nose-bleed, as it is apt to be painful and also bring about a feeling of fulness in the nose, and thereby increase the obstruction. Water well warmed, and containing an alkali combined with carbolic acid, may be employed; its value is greatly enhanced by the addition of glycerin.[1] Used in the form of a spray, it can be made quite efficient in dislodging the retained secretion, cleansing and purifying the passages. Unless the discharge is profuse and purulent, a spray of fluid cosmoline can be used to better advantage, often combined with some drug which has a healing action. This simple remedy frequently suffices to effect a cure.

4. *If a child snores, can it be prevented? Is it a sign of catarrh?*

This question may be answered yes, and no. If a child snores, this may and often does arise from obstruction to nasal breathing produced by nasal occlusion in various parts of the canals. It also very frequently arises from enlargement of the tonsils,—probably the next most common condition causing mouth-breathing and rendering snoring possible. There are some cases of snoring in children that do *not* arise from any diseased condition of the mucous tract, but appear to come on during profound sleep, such as is often noticed in adults. Snoring in adults, in many cases, does not depend upon any obstruction in the nose or throat; this I have often verified by careful inspection of subjects given to snoring. The prevention of snoring which is dependent upon disease will only be successful by removing the cause,—removal of all obstruction and the cure of mouth-breathing.

5. *Should a child that hawks or snores use douche or spray? If so, how often, and what to use with it?*

Very young children never *hawk;* the act of swallowing repeatedly or coughing takes the place of hawking, and the secretion in the fauces and post-nasal space is

---

[1] Dobell's solution, with a Davidson atomizer, No. 63.

generally swallowed, as in infants. A nasal douche should never be used for a child except in cases of dry or fetid catarrh, when it is of great use in expelling the retained secretion. A douche used under any other circumstances as a treatment for children is liable to do much harm to the middle ear and increase the cold in the head. As said before, a spray can be used with advantage, but only as often as is compulsory, say once or twice a day, and even then the mildest solution should be used,—one or two grains to the ounce, of boracic acid, chlorate of potassium, in glycerin and water, or, better, fluid cosmoline combined with borax or boracic acid. Medication of the nasal chambers can be very efficiently instituted by the use of the "Oliver Vaporizer," using fluid cosmoline or glycerin, to which may be added such drugs as are suited to the condition.

6. *Please give all the best methods for stopping nosebleed.*

Nose-bleeding occurs in the vast majority of cases in children from a superficial ulceration on some portion of the middle line of the nose, just within the opening of the nostril. This abrasion is caused from picking the nose, the finger-nail or handkerchief scraping off the outer layer of the mucous membrane. This superficial ulcer on the middle line remains unhealed for a long time, by reason of the free motion in the parts in blowing the nose, together with the low vitality of this thin mucous membrane covering the cartilage. Aside from injuries (falls, blows, and sharp substances thrust up the nostrils), the ulcer is the source of most of the nose-bleeds. Blowing the nose and wiping it with too much vigor, attacks of sneezing, cough, etc., also bring about the flow. Remembering that the seat of the bleeding is very low down in the nostril, and not high up, as is generally supposed, the ease with which the hemorrhage can be arrested by a little pressure will add greatly to the prompt and successful treatment of these attacks.

The finger (if child's nostril is sufficiently large) can be introduced and pressed up against the middle line of the nose for a sufficient length of time to stop the bleeding. Then, carefully withdrawing the finger and prohibiting blowing of the nose will often be sufficient. If the opening of the nostril is too small to allow the finger to be introduced, enough cotton, wool, or soft sponge can be pushed into the orifice to slightly distend the nostril, and pressure made against the outside of the nostril with the finger. If kept up long enough this will often succeed. If the bleeding has been great, or long continued, saturating the cotton with some astringent, such as a *weak* solution of alum, tannin, or tincture of iron, will cause arrest of the bleeding much more promptly. Probably the most satisfactory astringent to use on cotton is the watery extract of witch-hazel, familiarly known as "Pond's Extract." Saturating the piece of absorbent cotton or sponge and squeezing out about one-half, then introducing into the nostril, will be immediately successful. If there should be too much nervousness to try the introduction of any of these substances, a simpler plan is to apply ice or iced water to the nose, forehead, or back of neck at short intervals. Cold water snuffed up gently will often arrest the hemorrhage, or, if the child is too young to snuff up fluid, it can be squirted up the nostril with a small syringe. If urgent, extract of witch-hazel had better be used, and it will be with the happiest effect. It can be used full strength, but is painful, and had better be diluted with water, one-half or two-thirds in very young children, and in proportion in older ones. The witchhazel is in most cases entirely satisfactory, but if, after all these methods have been tried, the bleeding continues, a surgeon must be called and the nostrils plugged more perfectly. The position generally assumed in nose-bleed—that of holding the head downward over some vessel—is a very bad one and should be avoided.

The head should be held erect, and the blood allowed to flow into a towel, handkerchief, sponge, or absorbent cotton. To make a permanent cure, the ulcer—the cause of the bleeding—must be removed. This will have to be done by one sufficiently skilled to examine the nose carefully and treat the parts intelligently. The ulcer is often slow in getting well unless carefully managed. If the nose-bleeding arises from some other cause, or during an attack of acute illness, the nose should be promptly examined with a good light and the cause discovered and removed by the surgeon. At times, in children of plethoric habit, a small bleeding at the nose need not make the parent feel anxious at all, but it may be looked upon as simply an attempt of nature to get rid of too much pressure in the blood-vessels.

CHAPTER XII.

DISEASES OF THE EAR AND EYE.

Diseases of the Ear in Infancy and Childhood—The Care of the Ear in Childhood—Diseases of the Eye in the New-born—Treatment of Simple Ophthalmia—Contagious Ophthalmia; how to prevent it; its Treatment and Nursing.

For this chapter I am very much indebted to Dr. Charles S. Turnbull, of this city.

The symptoms of earache in a babe would lead those who are used to children to suppose that something irritated and troubled the child. Except slight fever, more or less restlessness, rolling of the head from side to side, perhaps backward burrowing into the pillow, nothing is noticed until the attendant reports a "running ear." The drum, or middle ear, fills with a watery fluid (the watery part of the blood, serum), or it may be pure blood, and this subsequently breaks through the drum-membrane and escapes from the ear. Often the perforation heals overnight, and apart from the staining of the external ear from blood, or the coating with a whitish residue from the serum, nothing is noticed. The perforation, however, does not always close. Discharges from the ear—caused by "taking cold"—behave like and may be compared to the discharge from the nostrils under similar circumstances. The secretion is first water, then mucus, then thick, stringy mucus, then matter, and last of all badly-smelling matter. Such discharge from an ear should only be carefully wiped out with plain absorbent, or borated absorbent cotton. When the discharge is odorous the doctor must be called.

If a babe seems to have earache, until the doctor arrives have resort to the hot foot-bath, and use dry heat to the ear; the "hot-water bag" or hot salt in flannel bag. The doctor will prescribe the necessary medicines.

We strenuously object to the use of drops of any kind, or water dropped or syringed into the ear. Fluids so used make an ocular examination of the ear, on the part of the doctor, quite impossible, and syringing macerates the parts and retards the subsequent healing process. Healthy babes' ears always get well if they are kept clean and not meddled with, so recollect that sins against the ears of children are more usually those of commission than of omission. Teething is the most frequent cause of earache and running ears. Water and soap we consider poisonous to the ears, and in answer to the query, "When should running ears be syringed?" we would say, "Never."

In syringing ears, except for wax or foreign bodies, there is great danger, especially to children, from impure water, varied temperature, shortness of auditory canal, peculiar susceptibility of parts from anatomical relations, and contiguity of brain.

DISEASES OF THE EAR AND EYE. 137

Discharges from the ear, in our opinion, unless they are sour or fetid, may be regarded as harmless in so far as the hearing is concerned.

The most essential point in the cleansing of discharging ears is the thorough freeing of the inside of the ear from secretion. This must be done by teaching adults Valsalva's experiment,—making forcible expirations while holding the nose and keeping the mouth shut.

Now, how can this all-important process be accomplished in the case of children? In only one way: by forcible inflation, or by having them blow their noses. *Every child should be taught, if possible, to blow its nose.* If it has not been or cannot be taught to free the nostrils of mucus, its chances of retaining hearing power in case of disease of the ear are much poorer than in the case of one who has learned to blow its nose.

The great secret in the successful treatment of all discharges from the ears is the recognition of the fact that so long as a discharge is not allowed to ferment, it will not become fetid, seldom even purulent.

As soon as a discharge from the ear makes its appearance, something is invariably dropped into it, or, in conformity with custom, it must be "syringed out." Nothing is more damaging to the successful termination of such cases. We would not complain if pure, warm water were used to syringe the ear, or, still better, if a salty or an alkaline solution were used, but the "Castile soap" is invariably added. Perhaps warm milk may be used. The soapsuds make an irritating solution, the milk one that rapidly ferments and becomes acid, so that the auditory canal—it may be a warm cavity filled with simple mucus or, perhaps, serum—is converted into one that is inflamed and filled with fermenting fluid. What next is done? A fluffy piece of cotton is rolled into a dense mass and stuffed into the auditory canal. The sour, fermenting, perhaps fetid mass, now corked up, fairly boils, and, where a harmless inflammation was in existence, an active and a dangerous one has certainly been started.

To remedy affections of the ear, general surgery has done but little, so that in many instances medical men are glad to get rid of "patients with running ears;" and this added to the prejudices in the minds of the community at large, and in some of the profession, too, as to the injurious effect of healing, or "drying up," as it is termed, discharges from the ear, has caused this affection, through ignorance or apathy, to be much neglected. We cannot in this connection omit a quotation from Saunders,[1] who tersely ventures a question or two concerning those prejudices which even to-day, alas! are urged against the cure of running ears.

"What argument can be assigned against the cure of this disease that is not equally conclusive against all others? Is any one an abettor of the obsolete humoral pathology? He will contend that the stoppage of a drain which nature has established is pernicious, and the morbid matter will be determined on the internal parts; but how can such a person venture on the treatment of any disease, even the healing of a common ulcer? Some years ago I thought this absurd doctrine had been totally exploded, and yet I constantly hear it adduced to deter parties from interfering with this disease. Is a child a subject of it,—the parent is told it is best to leave it to nature, and the child will outgrow it. Is it an adult,—some other subterfuge, equally futile, is employed. The truth is, the disease is always tedious and difficult, and *not* always curable, and many are disinclined to embarrass themselves with the case, who have not candor to make the true statement."

We have often been met with the objection,—and,

---

[1] "The Anatomy of the Human Ear," etc. John C. Saunders, M.D., London, 1806.

we must confess, it is generally well put,—" Why dry up the purulent discharge from the ear, since when suppuration is actually taking place, patients, as a rule, hear best?" True enough, as the discharge ceases in a case of discharging ear, hearing power is, as a rule, materially diminished; but we always make reply by asking the question, " Which is better, half a loaf or no loaf?" that is, to stop the discharge and save some hearing, *which will be permanent*, or allow the discharge to continue, and, in a greater or less time, lose all hearing? Then, too, if only for the abolition of the disgusting fetor which accompanies such cases, if for nothing else, it is well worth while risking any fancied extension of the inflammatory process. Children with " running ears" are tabooed by their class and playmates. Adults are tolerated, while, self-conscious of the sickening odor from their ears, they shun society, and imagine, not without good cause, that every one is aware of their infirmity. The majority of such patients are generally willing to forego the greater or less amount of loss of hearing power, if the offensive discharge can be prevented.

Beyond a doubt a discharging ear is a thorn in the flesh, to be withdrawn in the shortest possible time. Apart from the risk of damage, from a chronic discharge from the ear, to the hearing and subsequent happiness of the individual, is the undoubtedly compromised condition of the unfortunate's health. Fetid discharges run down into the throat, and poison the system. This is no fancy deduction, but a fact. By simple cleansing of the ears and teaching our patients how by Valsalva's method to blow the pus outward from the ear into the canal, which we keep thoroughly cleansed, we have met with most pleasing results in these same children, who, from pale and emaciated subjects, have grown into fat and ruddy specimens of humanity.

Again, we are often confronted with the objection that "if these discharges from the ear should be stopped, the disease will go to the brain." How did this idea originate? Because heretofore such heroic measures were used to check the discharge, because such caustic solutions were poured, and powders were insufflated, into the ears; furthermore, because no intelligent treatment was employed. In the majority of cases no careful ocular inspection of the parts was ever made, and extension of inflammation and disease to the inner ear, or even brain, resulted.

The majority of cases of running ears in children under two years of age would recover, and the hearing would not be damaged, if they were simply let alone.

Now, this statement may seem startling, but it is nevertheless true. Ordinary cleanliness is all that is necessary for the proper management of such cases. In the use of medicated solutions to be dropped into the ears of children, the anatomy of the parts must be understood. The auditory canal is short and the Eustachian tubes are patulous, and the solutions dropped into the discharging ears of children run directly into the throat. For this reason, if for no other, the syringe should not be used.

Admitted that nothing but pure water has been used, even this is too irritating for the ear, Eustachian tubes, and fauces (it must not be forgotten that the mucous membrane or lining of the drum or middle ear serves the purpose of periosteum, the same membrane that covers and nourishes the bone). Those who have accidentally gotten water up the nose will recall its unpleasant irritating effect.

After scarlet fever, measles, chicken-pox, bronchitis, etc., dentition or cutting teeth is the most fruitful source of running ears, and the tendency of all such cases is to recover, without damage to hearing, providing they be kept clean and nature be given a chance. How should the ears be kept clean?

We want to impress upon parents, and those who are to advise them, the necessity of using the utmost care in the art of cleansing the ears of children. Wax (cerumen), with which nature has furnished the auditory canal, is usually swabbed out weekly, if not oftener, with a twisted-up corner of a towel, handkerchief, or wash-rag soaked with water or soapsuds; and, more frequently than is supposed, a pin or the hair-pin is called into requisition. By these means the wax is pushed in and well rammed down, layer after layer, and at each washing a layer of desquamating epidermis is added (as is the cow's hair to the mortar), and this serves to bind the mass together and make its removal more difficult.

Masses of wax or dried skin, in fact, all sorts of foreign substances, pushed into the ear by unsuccessful attempts at cleansing with the wash-rag, etc., or foreign bodies designedly placed in the ears by children, are often the cause of a most distressing cough, for which the little patient is mercilessly dosed. To the experienced the peculiar spasmodic ear-cough can be recognized, and observant children often complain of a sticking or irritation below and behind the angle of the jaw, which symptom—and perhaps accompanying deafness—must point to the ear as a cause.

Children naturally rebel, and interference with their ears is generally a cause for war in the nursery; and it is just here that we wish to put in a plea for the juveniles, and condemn the usual practices of the best-intentioned of nurses and mothers.

With but a few exceptions, impacted wax in adults is found only in the ears of those who vigorously use water, soap and water, or wet cloths, to cleanse their ears from what they call dirt, and what we must recognize as absolutely essential to perfect hearing and a healthy condition of the ears.

Impacted wax must be first soaked by the instillation of a warm alkaline solution, and then can always be safely removed by syringing with warm water, which procedure is the only one in which we consider the use of water permissible; even here, however, had water not been injudiciously used in the first place, the wax would never have become packed.

In case it becomes necessary, for superfluous wax or the lodgement of dust, to wipe out the meatus, it should be done with a dry, soft cloth, or a damp towel.

About foreign bodies in the ear.—Children seem especially possessed with a suicidal mania for placing buttons, beads, seeds, pebbles, etc., into their ears, and the majority of doctors are prone to attempt to spoon, gouge, or dig them out. Whosoever attempts the removal of any foreign body from the ear of a child by other means than the syringe and warm water, and, if the child be frightened, without the use of an anæsthetic, certainly shows great want of experience, or heed to the warning of those who can advise.

Now that we have the petroleum products vaseline, cosmoline, and fluid cosmoline, all of which have the charming recommendation of never fermenting, the oils usually distilled must be omitted, because such vegetable products as olive and almond oils act sooner or later as irritants.

We never use anything else to drop into the ear, except, for larger children and adults, the following, which we would call "earache drops" (the aromatics contained preventing the small proportion of almond oil from fermenting):

℞ Cocaine muriate, two grains;
"Baume Tranquille," three drachms.•

We direct a few drops into the ear with glass dropper, previously dipped in hot water.

Babes should always wear an under flannel cap, and

larger children should wear, during the cold weather, caps which may be made of any warm material to suit the taste, but they should tie under the chin and protect the ears.

Too frequently warm caps are worn during the week, while on Sunday, as well as on high days and holidays, a gorgeous hat is substituted which leaves the head unprotected.

Especially necessary is the warm cap for a child who has suffered from ear-diseases. *On no account should cotton ever be worn in the ears.* Why? Because it acts like a cork and prevents the ventilation of the ears, and acts only as an irritant. If a discharge be present, the plug will induce fermentation and offensive odor.

A child with a running ear can go out of doors in cold weather. Unless there be actual pain present or commencing inflammation of the ears, a child should be warmly dressed and sent out just as usual. Housing of children with running ears is more apt, by engendering ill health and disordered digestion, to do harm rather than good. If the *feet are kept dry and warm*, fresh dry air is to be specially recommended with outdoor exercise.

Look upon water and glycerin and the much-extolled and detestable Castile soap solutions as irritant poisons, and taboo them, at our recommendation, as worse than useless. So also must we protest against the use of oils other than vaseline or cosmoline, because they become rancid.

In conclusion, allow us to say,—

For *collections of wax* in the ears, soak and syringe.

For *pain* in the ears, use dry heat and anodynes in full doses.

For *children's earaches*, never forget the hot foot-bath and aconite. Tincture of iodine behind the ears is less annoying, and does just as well for counter-irritation as a blister.

Wipe out *running ears* with absorbent cotton, and do not meddle too much with new cases.

Tickling under the angle of the jaw on either side, with sudden impairment of hearing, means irritation, and calls for special treatment.

Old cases of *running ears* call for an aurist's advice. *Never plug* discharging ears with cotton. *Never pick* the ears with anything smaller than the finger.

Given a case of deafness in an adult, insist on Valsalva's experiment. Given a case of deafness in a child, insist upon its frequently blowing its nose or else forcibly inflate, for it, the middle ears.

We have had particular reference to diseases of the ear in children, and have confined ourselves to the study of the hygiene and practical common-sense treatment such as experience leads us to commend.

In prescribing for the ear entirely too much guess-work is indulged in, and by waiting, too much precious time is lost.

It is recklessness—and this is putting it very mildly —to prescribe for an ear without first having made a careful examination of it, and he who orders anything to be dropped into the ear before making an ocular examination of the canal is not to be trusted.

DISEASES OF THE EYES.

Let us consider the nursing of diseases of the eyes in infants and children.

To the laity almost all inflammations of the eyes of new-born babes are known as "ophthalmia," but we want to specify and say that physicians, in a general way, recognize two forms. One caused by cold—the harmless variety—can be easily cured by careful cleansing and the use of astringent washes, such as rose-water or alum or borax (one grain to the ounce). The other, a most dangerous and easily contracted disease, threatens blindness. This latter form is not apt to spread among careful people, at their homes, but at a public institution, unless isolated at once, will be sure to sweep through and attack every child in it.

If asked how a baby's eyes should be treated immediately after birth, we would say, certainly not to a dose of soapsuds in the—to be condemned—popular primary scrubbing with which many a poor babe is tortured. The bad colds (snuffles) which attack such over-scrubbed babes affect the eyes, as well as the nose and throat, and often alarm the medical attendant and worry the mother and nurse.

If, as in such cases, the eye looks red and the secretion be gummy and sticky, like the white of an egg, or if a yellowish-white discharge be present, gathering in the corner of the eye, or perhaps gumming the lids together, a soothing eye-wash composed of

R Borax, five grains;
Paregoric, one teaspoonful;
Infusion of sassafras-pith, eight tablespoonfuls,

should be frequently applied with a soft piece of old linen, and about twice day a few drops may be allowed to run between the open lids.

Should, however, the eyelids, within the first twenty-four hours, puff up and swell so that the eye cannot be opened, and particularly should the secretion oozing from between the lids be creamy and of a yellowish, pinkish, or greenish color, then look out; the dread ophthalmia has started.

This disease, which contributes so largely to the blind in our asylums, calls for the most heroic treatment on the part of the medical attendant, and eternal vigilance on the part of the nurse. At this juncture both physician and nurse must incessantly and conservatively, yet resolutely, attack the disease, and unceasingly fight night and day.

The swelling and inflammation must be combated by

cold applications, day and night, made by means of small pledgets of linen, which are lifted cold and wet from a block of ice and laid upon the eye, but be careful that they cover no more than the burning eyelids. These pledgets will be required to be renewed frequently, at intervals of from fifteen to twenty minutes. The lids at the same time must be gently separated, and the discharge allowed to escape, or be carefully wiped away with pieces of absorbent cotton dipped in fresh water or salt or borax to make a weak solution.

If the doctor be willing to personally supervise the daily preparation of a gallon or more of corrosive-sublimate wash (1 part to 6000), nothing could be better. Sponges are dangerous. Use absorbent cotton, as it is clean and can easily be disposed of.

Great care must be exercised that the cold applications be not kept up too long,—*i.e.*, without intervals of a few minutes' rest, say fifteen minutes every two hours. Be careful also about the ears, and see that no water trickles into them; also see that the hair be kept dry, and the pillow as well; and be most particular that the patient's and attendants' hands are kept clean and never put near the face, always regarding the discharge and every sort of eye-wash used, as rank poison.

So long as the discharge be creamy it must be considered as corrosive (like acid or vitriol); and it is during the first stage, when the swelling is so great that the lids cannot be opened, that the following injection, to be repeated every three hours, must be gently squirted (with a round-ended medicine dropper) from the inner (nasal) angle of the fissure between the eyelids:

℞ Sulphate of morphia, two grains;
 Chloride of zinc, two grains;
 Rose-water, ten drops;
 Distilled water, five tablespoonfuls.

Now, recollect such eyes cannot be watched and

cleansed and treated up to bedtime, and then neglected because the baby or its attendants sleep. Oh, no, lest the little one's eyes melt away during the night. Long naps are not desirable, either in the case of children or nurses, as a good night's rest has cost many an eye. We repeat, the treatment must be kept up night and day. We have said the secretion is corrosive in its action upon the structures of the child's eyes, and as the eyeball, from pressure of swelling, is damaged, the discharge is doubly fatal. Then, too, the child might turn in bed, and the discharge, by gravity, would run into and inoculate the fellow-eye. Yes, this is not all; it is deadly poison when transplanted to any other person's eye, and nurses, mothers, and medical attendants cannot be too careful. Even the poor laundress often becomes a victim, and this poison has too often been the cause of the loss of valuable eyes, to which, through ignorance or carelessness, it has, by touch only, been conveyed. Keep the poor little one's excoriated cheeks and skin, worn out by washing, well anointed with vaseline. So also the edges of eyelids, nose, and nostrils. Keep pledgets of cotton (not absorbent cotton) in the ears. Keep the strength up, and give cooling medicines such as the usual fever mixtures, nitre, quinine, and aconite.

The baby's sleep will be greatly disturbed, its nervous system racked, it may take cold, it may become sick, it may die from irritation and exposure. Take the risk, and if you falter in accepting the odds, think of the heart-rending appeal of a pair of sightless orbs. Think that the flaxen-haired girl is to be a "blind Nydia," or the boy a dependent, helpless man. We are emphatic, because we have seen the deplorable ravages of the disease we have but partially described, so we urgently warn you to immediately care for it, and promptly seek counsel and experience to fight this dreadful disease. It blinds individuals in families, it blinds our children in our asylums, and two and three at a time; it blinds grand nurses, loving, faithful Sisters of Charity, and even prominent and skilful physicians.

Unless a physician be at hand to apply such local remedies himself, we consider it criminal to recommend or order caustic solutions to be dropped into the eyes. The physician only can turn the lids, brush over any caustic solution necessary, and immediately (for fear of over-effect) wash it off again; but stronger applications than those we have named are not to be recommended for home use.

We have only referred to the treatment of individual cases at home. In asylums, where infants are brought in daily, we insist upon immediate isolation of the child as well as its attendants and their wash. If the disease be progressing to a favorable termination the discharge will become less, the swelling of the lids will diminish, and the child will be able to open its eyes. At this stage we commence the use of a slightly stimulating salve:

R Yellow oxide of mercury, one grain;
Vaseline, three drachms.

To be rubbed on the edges of the eyelids about three or four times a day, or just before the babe goes to sleep. Beyond this we cannot advise any other treatment save that which your medical attendant may order.

We would in this disorder make an exception, and say that without medical advice and guidance "ophthalmia in the new-born" must, as a rule, mean blindness, whether of one or both eyes. In the contagious ophthalmia of asylums, where one sore-eyed baby poisons the rest and where simple, yet decided local treatment is required, we particularly commend a peculiar compound called fifty per cent. "boro-glyceride." This is made by boiling together boric acid and glycerin— sixty-two parts of the former to ninety-two parts of the latter—until the product loses weight and weighs but one hundred parts. This in cooling resembles in consistency and appearance ice or "glacial phosphoric" acid, and is found to be very hygroscopic. To dilute it, glycerin must be employed, and the best method for its preparation is, when freshly made, to add to it glycerin in such proportion as to make a fifty-per-cent. solution. This makes a preparation of the consistency of honey, to which can be added iodine, tannin, resorcin, carbolic acid, iodoform, morphia, atropia, eserine, etc., as may be desired. The ointment of boro-glyceride is made after the following formula:

℞ Sol. boro-glyceride, fifty per cent.;
Vaseline, six drachms;
Oil of rose, q. s.

This makes a thoroughly stable ointment which neither becomes granular nor precipitates the boric acid. We feel confident that in this compound we have a most valuable remedy, and when called upon to combat the appalling epidemic, such as often occurs in our asylums, we feel much more secure—especially in the case of young children—when using this excellent remedy. The fifty per cent. boro-glyceride, on account of the great affinity of the solution for water, and the rapidity with which it absorbs it and liberates the finely subdivided particles of boric acid, not only acts as an astringent but also as an antiseptic. We believe it is just the substance that we are in need of for the treatment of all forms of chronic inflammations of the eye, especially "contagious ophthalmia."

The use of the ointment mentioned we consider an essential part of the boro-glyceride treatment, and it must be continued for at least two months after all discharge has ceased.

This disease, like every form of ophthalmia, is only

transmitted by inoculation, and under proper care need not be transmitted to the fellow-eye.

What is proper care? Proper care is two skilled and trusty nurses, the one for the day, the other for the night, *who never leave the patient;* or, as an extra precaution, better say four, so that one of the two on duty will be sure to keep awake.

As soon as the little sufferer shows the least tendency to open its eyes, it should be encouraged in its endeavors. Darken the room moderately, so that the influence of bright light does not make it shrink. The opening of the eye is beneficial in two ways,—the movements of the lids work the corrosive secretion out from between the eyelids, and they stimulate the circulation in the affected parts.

When a child opens its eyes the danger is over, only a relapse must not be allowed to occur. The iced applications have only to be continued until the swelling of the lids and the creamy character of the discharge have disappeared. No child need lose its eyes from ophthalmia, and no child does, if faithfully treated in the way just described.

Knapp says he is convinced that nothing is so powerful in diminishing the violence of this dreadful inflammation as cold, and he is afraid that warmth may temporarily increase it and favor destruction of the eye; and says furthermore, among all questions in ophthalmology—that of cataract, perhaps, excepted—there is none so important as the treatment of "contagious ophthalmia."

CHAPTER XIII.
DISEASES OF THE THROAT AND AIR-PASSAGES.

Croup and Diphtheria—Simple Spasmodic Croup, what its Symptoms are and how to treat it—Membranous Croup, its Symptoms and Treatment—The Difference between Membranous Croup and Diphtheria—Diphtheria as a Cause of Membranous Croup—The Nursing, and the Use of Household Remedies in their Treatment.

THE diseases which probably cause the most alarm among mothers, and justly so, are those which affect the throat; especially when there is the slightest possibility of the disorder being either diphtheria, or membranous croup. I propose to give in as simple language as possible, so that it can be readily understood, a description of these various acute diseases of the throat in children, not for the purpose of enabling the mother or nurse to make a diagnosis,—that is not her business,—but simply to enable her to carry out thoroughly, conscientiously, and intelligently the nursing, which forms so important a part in the treatment of these diseases. Undoubtedly, in times gone by, children suffered very much more from croup than they do at present. The change in this respect has been brought about by the doing away with the short-sleeved and low-necked dresses of children, and the thorough understanding of the fact that in a climate as changeable as ours the whole surface of the body, from the neck to the feet, should be protected from the sudden chilling of the surface by wearing a garment, be it ever so thin, of either wool, or wool and silk mixed.

We hear a great deal of diphtheria, and it is well for us to understand what is meant by this term. The word has been applied to a disease which is characterized by a deposit upon the throat, and, consequently, every time a deposit is noticed upon a child's throat it is at once thought to have diphtheria. This is a mistake; what really constitutes this dreaded disease is a profound constitutional poisoning, caused by exposure to sewer-air, polluted water or milk, exposure from contagion of the same disease, by which the system is profoundly poisoned with the local trouble, manifesting itself in the throat, larynx, or air-passages. A child may have diphtheria without any appearance of membrane in its throat; then again deposits may be found on the tonsils, as in quinsy, which is not diphtheria at all; but for the sake of caution, children affected with sore throats should be isolated until seen by the doctor. Those who are well should not use the same spoon or drinking-cup as those that are ailing; this applies to adults as well as to children.

There are mild forms of diphtheria, affecting adults, during which the constitution is not sufficiently affected to prevent their going about their daily avocations, and such individuals are beyond a doubt the means of carrying contagion to those with whom their breath comes in contact. There is no doubt but that many times diphtheria, and other diseases even worse, have been carried to children by the foolish and useless practice of kissing to which the poor little ones are subjected. It seems strange that a child who will be protected most energetically from draughts, from every form of open contamination, will be permitted to come in direct contact --the most thorough means of propagating contagion --with diphtheria, whooping-cough, scarlet fever, and worse, by kissing. I dwell upon this, because I feel that parents should instruct their nurses to prevent their children being kissed by other children and by strangers.

A sore throat should always be looked upon with suspicion in a child, and carefully treated; and especially if the child's neck is swollen on the outside, the glands enlarged, should there be a coated tongue, offensive breath, and great prostration. No one should attempt to treat a case of this kind without immediately consulting the doctor; it is a recognized fact that should by chance the little patient have been exposed to sewer-air from a defective stationary washstand in the bedroom, or from the bath-room or closet, a simple catarrh of the throat from cold will allow the diphtheritic germ, or whatever the poison may be, to be engrafted upon it or be absorbed into the system, and the only means of counteracting its influence would be the most active cauterization of the throat, and placing the child upon stimulants and medications that will strengthen it.

Whenever a child complains of a sore throat and there seems to be redness of the throat inside, or the tonsils become swollen and the child complains of swallowing, has a cough, it is always safe and indeed a proper thing to spray the throat and nose with some mild solution that will at once cleanse and soothe it. This will never interfere with the doctor's treatment, and can be done before the doctor comes. The best forms of atomizers that I know of, and one of which should be in every household, are the Davidson's small hand atomizer, which is valuable for spraying the throat and nose, and that of Oliver, a more expensive one, but the very best so far made for inhaling lime-water or such substances as are used in cases of croup. A solution known as Dobell's solution can be used in these cases; also in cases of whooping-cough, or in all forms of sore throat.

There are many diseases of children that are accompanied by sore throat or sore mouth, that require constant cleansing of the mucous membrane, or the deposit, or mucus, will accumulate, become offensive, or finally, as in diphtheria, become putrid. Great care should be paid to the slightest throat-ailments of children, because an irritated throat is often an opening for the entrance of infectious disease.

A child complaining of difficulty of swallowing should under no circumstances be allowed to go out of doors, especially if the weather is damp or at all raw or cold. The throat should be gently sprayed with the solution just mentioned, and some counter-irritation made to the outside, using St. John Long Liniment, or simple rubbing with vaseline, sweet oil, or cod-liver oil. If the child's voice is slightly muffled in speaking, if it speaks through its nose, should there be tenderness upon pressure and slight enlargement of the gland under the angle of the jaw, on one side, and the attack come on suddenly, in an older child with a chill or high fever, the probability is the attack is one affecting the tonsils,—one of quinsy. The household remedies to be used in such a case would be a fever-drink of one teaspoonful of spirits of Mindererus (liquor ammoniæ acetatis), one teaspoonful of sweet spirits of nitre to half a glass of water, sipped at frequent intervals, for a child about five years of age. The feet should be put in a hot mustard-bath and the stockings kept on afterwards, the throat sprayed, the child allowed small pieces of cracked ice; the food for at least twenty-four hours should consist of beef-tea or chicken-broth, or probably still better, if the child will take it, milk and lime-water, or Vichy water, half and half. In all these acute throat-troubles of children I have found the early application of a handkerchief wrung out of cool water, tied round the throat like a cravat and covered by oiled silk, of very great service in sore throats that come from exposure to dampness, getting the feet wet, or where the voice is husky with a slight cough, which has a tendency to become dry or ringing, especially towards night. This treatment will often avert an attack of croup. Throat-troubles in children are by far too serious to allow of any time to elapse before receiving the most thorough treatment; when a child complains, it is far better to send at once for the doctor.

CROUP.

There are some children who are particularly liable to croup : it seems a family characteristic. Boys are more apt to have it than girls,—possibly because they are more exposed; indeed, the tendency to "harden" children by letting them go bare-legged, or with low necks and short sleeves, will sooner or later bring either a bad attack of croup, or bronchitis, as a consequence. Children who go permanently, as do many of the poor, without shoes or stockings, are not as apt to contract these catarrhal diseases as those who temporarily have their feet exposed to cold, sudden changes, or dampness ; but the child who is over-dressed will perspire and be as liable to attacks of croup or bronchitis as one who is insufficiently dressed. What is ordinarily known as croup is that affection which comes on suddenly at night, accompanied by a dry, ringing cough, difficulty in breathing, all evidences of threatened suffocation, without any marked previous symptom. It usually occurs at night. The child will wake from sleep with all the symptoms that are most terrifying. Its evident spasmodic character has given it the name of spasmodic croup. It may be the result of cold or of an overloaded stomach, and children who are subject to this affection should not be allowed heavy suppers.

Although the breathing is greatly interfered with during this attack, the sounds are ringing, the cough sonorous and brassy and loud ; this is an important matter, as it shows there is no deposit of membrane to muffle the sound. The treatment of such a case should be as follows : a sponge wrung out in water as hot as can be borne by the child should at once be tied round the throat, and kept there by a towel, or better, oiled silk; the feet and legs as far as the knees should be immersed in a hot bath containing a few tablespoonfuls of mustard flour, and be kept in the water for at least fifteen minutes, then thoroughly dried, and a pair of stockings put on. The child should be given a half-teaspoonful or a whole teaspoonful of syrup of ipecac,—the latter if the child is over four years of age,—followed by a drink of water every fifteen minutes until it vomits; after which the spasms will cease, and the child will turn over from exhaustion, and sleep the remainder of the night. A powder of half sugar and half alum, given in teaspoonful doses or given with the ipecac, will hasten its emetic action. If the child is anxious and restless after the attack of croup has subsided, sleep seems impossible, or some cough still remains,—for a child from a year and a half to two years old, ten drops of paregoric in a teaspoonful of glycerin, repeated in an hour, will have a quieting effect, or ten drops of sweet spirits of nitre, given in a little sugar and water, can be given every half-hour or hour until the child is quiet. If the bowels have not been moved during the day, or are constipated, an injection of warm suds will frequently bring relief.

Possibly after a good sleep the next day the child will be in ordinary good health without any marked evidence of exhaustion from the attack of the previous night; or it may be slightly droopy, have an occasional cough which is rather ringing; the appetite may be impaired. If such is the case, it would be well to give the child a dose of castor oil, and by all means keep it in the nursery. If the child is over a year old, give it from five to ten, or if the child is three years old, fifteen, drops of the aromatic spirits of ammonia, every three hours. If children who have a tendency to hoarseness and croup have their feet washed in a basin of cool water every night before retiring, they will often escape an attack. It should

DISEASES OF THE THROAT AND AIR-PASSAGES. 157

also be borne in mind that it is not essential that a child should have a bath every day, as some mothers believe; if the weather is damp or raw, the child at all droopy, the surface can be very quickly sponged, simply for cleansing, and with salt water, and the circulation increased by a thorough rubbing with a soft towel. I think if the rule, which seems to be a cast-iron one with many mothers, of bathing children every day and taking them out in all kinds of weather, could be occasionally broken, we would not have so many sick children.

MEMBRANOUS AND DIPHTHERITIC CROUP.

There is nothing more appalling, or none that seems more strenuously to resist all efforts at successful medical treatment, than that form of croup which is accompanied by the presence of thick mucus in the air-passages of the larynx, or diphtheritic membrane. When speaking of diphtheria I dwelt upon the fact that it was a constitutional blood-poisoning, which could exist without the presence of membrane in the throat, and also that the membrane, when it did exist, need not be limited to the portions of the throat that we can see, but could extend to the whole surface of the mucous membrane down to the lungs, and upward into the nose; indeed, when it does so, it clings firmly to the mucous membrane, leaves a raw surface when detached, and if allowed to remain acts as an obstruction to the entrance of air, or decomposes, becomes putrid, and in this way acts still further as a poison. It is on this account that the spraying of the throat with an alkaline, antiseptic solution, which softens the membrane and prevents its decomposition, has been recommended in the most trivial throat-complaints as a precautionary measure.

Membranous croup is recognized to be of two varieties,—one the result of catarrh, when the obstruction is caused by very thick mucus in the larynx which may become tenacious like membrane, and the other due to the deposit of diphtheritic membrane, which may be found in connection with the appearance of deposit in the other parts of the throat, or may be simply limited to the air-passages. A child with membranous croup will probably be ailing for a few days; its cough will at first be croupy, its voice will become husky, the cough will cease, or nearly so; the breathing will become labored, and finally will be plainly heard at a distance; the child will show by its movements that there is narrowing of the passage in its larynx; the nostrils will dilate with each inspiration; the child will breathe with its mouth open, it will desire to sit up, so as to get all the air possible; the lips will become blue, owing to the interference with the circulation; the child will become anxious, restless, appealing to every one around it for relief; it will refuse food, but will drink water in sips at a time. If the obstruction continues or increases, the extremities become cold, the pulse rapid and at times irregular; and finally, if nothing is done for the relief of the child, it will gradually die of asphyxia, or a sudden shutting-off of the supply of air.

A peculiarity of the breathing in obstruction of the larynx is the contraction, or sinking in, of the lower portions of the ribs, or pit of the stomach or epigastrium (in health these parts expand), and the marked depression, or sucking in, of the lower portion of the neck, just above the breast-bone. But probably eight cases out of ten of membranous croup with which one comes in contact are due to a diphtheritic blood-poisoning, which we may attribute, in our larger cities, to defective stationary washstands in the sleeping apartment or nursery, absolutely conveying the most poisonous form of sewer-air from a bad cesspool or drain directly to the infant's room. One should never rest assured that the trap of the bath-tub, or its overflow, is secure, or that of the stationary washstand in the child's nursery or sleeping apartment; these should be banished to an adjoining room, where they have free communication with the fresh air, and should be well supplied by some form of disinfectant, as that used by the Germicide Company. But then again a child with a slight sore throat may be exposed to the gases coming from an open sewer; or through the air supplied to a furnace, which is passed over putrefying matter, or stagnant water in the cellar; to the breath of one convalescent from the disease, or who has it in a mild form, communicated by kissing. In addition to this, milk when mixed with water which has been contaminated by sewerage, has been known to produce this disease; it is on this account that the catarrhs, especially of the croupy kind, however slight, should be attended to at once in children, as they constitute a point of entrance of the diphtheritic poison into the system; and as children over a year old are very much more subjected to the sudden changes that produce these catarrhs, all forms of membranous croup are more apt to take place at that time.

Let us dwell for a few moments upon the nursing of such cases. There are two facts most important to bear in mind as causes of death: the one, the interference with the entrance of air into the larynx or air-passages; the other, the poisoning of the system, and the weakening of the heart, which accompanies diphtheria, and makes the patient more readily succumb to the influences of deficient aeration.

If a child has a croupy cough during the day, rapidity of breathing, slight fever, no time should be lost to influence at once the mucous membrane, to establish secretion and give relief. The child should be kept in its nursery, and I may say that in all cases of bronchitis or pneumonia, and even croup when a child has to be lifted constantly from its bed, or cannot be comfortable for any length of time in one position, it should have a long flannel wrapper extending much below its feet, that can be opened in front and at the back, with high neck and long sleeves. It is absolutely necessary that the air of the nursery be not only pure but warm, the temperature about 80°. The air should be moist, and for this purpose some arrangement must be made to have a constant supply of steam; this can be done by placing a kettle on the stove if there is one in the room, or by using a tea-kettle with a lamp under it for that purpose. If the case seems to be one of more than an ordinary slight cold, no time should be lost in allowing the vapor of an alkali to come in contact with the mucous membrane, and there is none better for this purpose than that which comes from slaked lime. Lime can be placed in a bucket beneath or beside the child's crib, and the crib covered over with a sheet in the form of a tent, allowing the vapor to accumulate within; if the child is only satisfied to remain on the mother's lap, the vapor can readily be brought in contact by a sheet thrown over the bucket, and extending over her own shoulder. The advantages of the vapor from lime are its moisture, its warmth, its alkalinity; and if membrane should form in the throat lime-vapor honeycombs it, the same as lime-water does the curd of milk. If the child seems to suffer very much from the obstruction, that is to say the breathing becomes rasping, the spray from an atomizer—and a steam one is much to be preferred—should be used constantly; or if the child is old enough it can probably for several minutes at a time inhale the vapor from the Oliver vaporizer. The solution to be employed in these cases is one-half lime-water and one-half Dobell's solution, or if two vaporizers are in use, as should be the case, so as constantly to have some form of steam inhaled, one solution should be of turbid lime-water, made by adding a teaspoonful of *liquor*

DISEASES OF THE THROAT AND AIR-PASSAGES. 161

potassæ to two ounces of lime-water. These cases are so important as regards the very early treatment that I do not hesitate to suggest that the mother or nurse, if she cannot get the advice of a physician at once, should lose no time, but place the child upon a treatment as follows, until the doctor arrives.

If the child's croupy cough or hoarseness should continue during the daytime even after the hot foot-bath and the warm applications to the throat (flannel wrung out in hot vinegar and water, and covered by oiled silk), the following fever mixture, which will possibly also loosen the cough, should be given until the doctor's arrival:

Solution of Acetate of Ammonium, one ounce and a half;
Solution of Citrate of Potassium, one ounce;
Compound Syrup of Squill, one drachm;
Simple Elixir, sufficient to make three ounces.

Dessertspoonful in water every two hours, for a child of two years.

The question of an emetic may be a serious one if suffocation seems imminent. If the child struggles for breath, or its face becomes bluish, its extremities become cold, and the other signs of which I have spoken exhibit themselves, a half-teaspoonful of alum, with the same quantity of powdered sugar, or given with water, would form the safest emetic. A harsh emetic such as ipecac may disorder the stomach without a correspondingly good effect, indeed may increase the debility, and should not be given without the doctor's advice. In addition to this, at least every hour a child of over a year should take five drops of the aromatic spirits of ammonia, in water.

Warmth and moisture should be applied to the chest and throat. Some physicians recommend the use of some material that will readily absorb large quantities of liquid; a mass of lint or folds of flannel or some patent material may be connected by shoulder-straps and tapes at the side, soaked in warm water (wrung out to prevent its dripping), placed on the front and back, and covered with oiled silk, and changed about twice a day or oftener. Care should be taken when the change is made; it should be done by placing the hand beneath the wrapper, and without exposing the child's chest at all. Others prefer ordinary flaxseed poultices, made in the same way. My own preference is for two flannel bags filled with hops and quilted; these can readily be attached by safety-pins on the outside, and thus kept in place; oiled silk can be stitched on to each one, so as to prevent evaporation, and these bags can either be wrung out in water, or some soothing or stimulating liniment, as the occasion requires.

The child's diet must be carefully attended to. It should be encouraged to drink freely of milk, and lime- or Vichy water, chicken-broth or beef-tea, or, if it prefers it, small quantities of beef-juice, or beef extract or peptonoids. Wine-whey, in cases where there seems to be debility, is absolutely essential, or ten to fifteen drops of port wine, whiskey, or brandy, in water, or a teaspoonful of malt extract should be given every three or four hours.

In regard to the feeding of sick children, absolute regularity and accuracy must be observed in their diet. The food should be of the most nourishing character, requiring least digestion, and in as small bulk as possible. There should be a schedule kept of the exact amount, and time it is given, so that the doctor may judge of the fact as to whether the child needs more or less in the twenty-four hours. In some diseases, such as diphtheria and typhoid fever, the child's life depends on this systematic nourishment. In the latter, the dietetics form the most important consideration, because for a long period, at least three weeks, the child's life is to be sustained by a diet so regulated that it will give all possible nourishment, and at the same time not unduly stimulate the moving of the bowel, or act as an irritant to those portions which are already inflamed or ulcerated. In the nursing of diphtheria, the most important points to bear in mind are, first of all, that it is a disease capable of being greatly modified by careful treatment and nursing; that it is directly contagious by means of contact with the membrane; that systematic nourishment and stimulation are absolutely essential.

Indeed, all continued croupy coughs should be looked upon as serious, even should no evidences of diphtheria be detected in the throat. Recently the great facility of placing a tube through the mouth into the larynx has been demonstrated, with the result of a great saving of life. It is an operation devoid of the horrors of opening the trachea, and can be made use of very early in the case, before the child is exhausted and the air-passages are filled with membrane or thick mucus.

Whether due to diphtheria or not, the treatment and nursing of membranous croup are practically the same. A child suffering from this disease should have its throat washed and sprayed with antiseptic cleansing solutions frequently, in order to prevent decomposition of the membrane; the air that it breathes should be pure, and at the same time charged with those materials that soften the deposit of the membrane and prevent its putridity. For this we have the vapor of steam containing lime, the products of tar or carbolic acid, and chlorine; but as I have gone over this matter thoroughly when writing of scarlet fever, it will not be necessary to repeat it here.

CHAPTER XIV.

DIARRHŒA.

The Causes of Diarrhœa—Over-feeding; Tainted Milk; Decomposition of Food; Undigested Starches; Teething; Hot Weather—Inflammatory Diarrhœa—How it can be avoided—Change of Diet necessary, also Absolute Quiet, Pure Drinking-Water, and Fresh Air—The Character of the Diet—Importance of Peptonized Milk and Nutritious Injections.

THIS is one of the most important subjects for our consideration. We have it varying from simple frequency in the natural movement, to that most violent form which occurs in summer-time; but as the one leads to the other, mothers should be instructed how to treat the earliest symptoms, so as to check them before there is any necessity for medical treatment.

Indigestion is the one great common cause, and, as we have before noticed, indigestion is produced by *improper food, over-feeding,* or those depressing conditions which surround the child and prevent the digestion of its food, allowing it to ferment and act as an irritant. All these causes can act equally well in winter as in summer, at the sea-shore or mountains as in the crowded cities with unhealthy surroundings, in older children as well as in infants; but as a rule they are very much more active in the overheated atmosphere of our densely-populated city during the summer, and infants that are obliged to stay in town are much more depressed, have less vitality, and their food is more apt to be tainted by the germs of putrefaction.

Over-feeding is so constantly a cause of diarrhœa that it deserves a few words of caution. Many young mothers believe that if an infant cries it is always an

DIARRHŒA.

evidence of hunger; this is by no means the case; it may be simply thirsty. If it is nursed at its proper time and has received its usual amount, a few spoonfuls of water will often quiet it instead of having recourse to the breast or bottle. I cannot lay too much stress on the importance of water, especially in the summer-time; when given judiciously and frequently it may often save the child an attack of summer-complaint. Give cool water; do not give iced water, but let it be pure, filtered always, and if there is the least suspicion of its purity have it boiled. Do not put sugar in it. Mothers should watch carefully the urine of their children; if the child cries suddenly, in a rather shrill cry which is characteristic, it has probably wet itself and should be immediately changed, the parts thoroughly dried and slightly greased to prevent irritation. If there should be a reddish deposit on the diaper, this is an evidence of indigestion or want of water to dissolve these crystals of uric acid. In such cases the urine is usually scanty. It will be found that a teaspoonful of soda-mint solution in a small teacup of water, or lithia-water, which the child should drink as freely as it will or take it from a spoon, will soon, if kept up for two or three days, correct this condition.

The child will cry if its position becomes irksome. It requires to be turned over in bed if it feels the cold, and requires the warmth of the nurse's arms to lull it to sleep, if its clothing irritates it, if the bands of its clothes are too tight, owing to flatulence extending the abdomen. Frequently it will cry from irritation of the skin, due from what is known as prickly heat. A little vaseline rubbed on will soothe it; or if the cry be sudden and sharp, it will be noted to be one of pain. Frequently this will be caused by the passage of its water, and the doctor's attention should be called to the parts. All these matters should be always taken into consideration and carefully investigated before the child should receive more food than its regular nursing every two or three hours.

Over-feeding is generally recognized by the evidences of the insufficient digestion in the passages, or by the frequency of the movements, although they may not deviate from the normal condition; or the stomach may become overloaded, and the uneasiness of the child following its taking of food, its restlessness during sleep, possibly nausea or vomiting, will be the consequence. It is a fortunate thing that a child vomits very readily, and therefore food that disagrees with it, fermenting in its stomach, can be gotten rid of without passing through the intestines, which are so easily irritated by such matters. If these symptoms just noted be present, the child's breath heavy or sour, and especially if it is a bottle-fed baby, it will be well for the mother to encourage vomiting. The simplest means of so doing is to give the child a glass of water and follow this by giving a half-teaspoonful of syrup of ipecac every fifteen minutes until vomiting is produced.

The next cause of diarrhœa for us to study is that produced by the food. It is a well-known fact that the nursing mother, living on certain articles of diet, can so alter her milk as to make it a potent cause of indigestion. It is also known that emotions, passion, fright, will so alter the character of the milk as to produce the same effect. Also any irritation of the nipple producing in the child a sore mouth will be the starting-point of a catarrh of the stomach and bowel; this can be readily obviated by cleanliness, the use of borax, glycerin, and rose-water as a wash, and vaseline to the nipple to prevent cracking. I have frequently observed the fact that a mother can depart from this rule of careful dieting on many occasions without altering her milk, in a way that would affect her own child, whereas the same milk given to another child, should she act as a wet-nurse, would be followed by harmful results. The close sympathy between mother and child is thus kept up after its birth. All physicians have known of cases of infants at the breast that have suffered from violent attacks of diarrhœa, convulsions, the result of violent emotion, fright, or passion on the part of the wet-nurse. It is this, together with the high living to which they are unaccustomed, the sedentary occupation which encourages biliousness, carelessness in diet, the inability to watch them in every particular as to their functional health, morals, personal cleanliness, that renders wet-nursing so unreliable.

Certain substances have a peculiar tendency towards the milk. Castor oil taken by the nursing woman will purge the child; mercurials will sometimes have the same effect; onions, garlic, will be noticeable in the milk and cause indigestion. Alcohol will find its way to the milk; this fact has been taken advantage of by giving the mother malt extract, which not only aids her own digestion, but the alcohol which is in it in small quantities, and also the diastase or nutritive qualities of the grain, possibly stimulates the digestion of the child.

The one potent cause of intestinal disturbance in children, in summer, is stale cow's milk; and by the word stale I mean any milk that has stood some time after milking, unless it has been boiled or subjected to intense heat. My own experience is that milk fresh from the cow, that is warm, and from a cow that is healthy and clean and has been carefully fed, can be given to a child from a bottle without any manipulation, except slight dilution, and can be digested, whereas the same kind of milk after standing several hours might cause intestinal disorder. It is the atmosphere which does this; it contains germs of putrefaction, very much more potent in the summer, and abounding in our large cities. It is almost impossible to distinguish this tainted milk from that which is fresh and pure, the changes are so insidious.

We have said that the diarrhœa of infancy and childhood included every form, from simple looseness of the bowel to cholera infantum. The healthy passages should have little odor, no mucus, no curds; should have a consistency and appearance of prepared mustard. Occasionally from cold or slight indigestion, probably from sore mouth, one or two passages may change in appearance; they may become greenish, variegated in color; they may become more frequent, contain a few curds. If such be the case, and the child is nursed, a teaspoonful of sweet oil, with the same of solution of soda-mint, will often be sufficient to bring the passages to their normal condition. If the mother or wet-nurse show signs of biliousness, a dose of castor oil given to her will probably have a good effect upon the child. If the child is bottle-fed and curds still appear in the passages, the child is getting food too full of curd, or its digestion is weakened. The bottle should be omitted for one feeding, and some barley-water or weak chicken-tea be substituted, and the child's digestion stimulated by giving five or ten drops of the solution of lactopeptine or a grain of a reliable pepsin in a tablespoonful of water, with five drops of whiskey, before the next feeding; or as much of a powder containing equal parts of lactopeptine and subnitrate of bismuth as will cover (not heaped) a dime.

If the child does not seem desirous of food, under no circumstances press it; should it crave water, allow it to drink, not iced water; if there is the least irritability of the stomach, give a tablespoonful of lime-water or a teaspoonful of soda-mint solution, with about five drops of whiskey, every fifteen minutes or half-hour, until the stomach is settled and the child quiet.

The evidences of catarrh of the stomach and bowel are pain, restlessness, heat of body and head, with unnatural coldness of the extremities, or there may be fever. The character of the pain is peculiar, the child will cry out at intervals, draw up its knees; the abdomen will be tense, the muscles firm and rigid and distended by gas. The passages, in such a case, will be liquid usually, containing curds, lighter in color than normal, and have with them large quantities of slime (mucus). The tongue will be coated; the child will probably crave water if there is much heat or dryness of skin. Now, this condition can be brought about at any time of year from exposure to cold, or from irritating food, but it is usually found in the summer in bottle-fed babies. Should it be allowed to continue for any length of time and the cause not removed,—by this we mean the child still subjected to the intense city heat and the same bottle-food,—the diarrhœa will soon become excessive, evidences of inflammation will appear: vomiting, intense thirst, a diarrhœa which is almost water, tinged with green and containing minute specks of greenish matter, finally entire loss of control over the bowels, the child will gradually lose consciousness and die of exhaustion.

True cholera infantum is not a very common disease, and what is ordinarily known as cholera infantum, such as I have described, is simply the aggravation of an ordinary simple diarrhœa which has been improperly treated at the commencement, either through ignorance or the force of circumstances. It is impossible for one to so guard a child against the many causes that conspire to produce catarrhal affections; the change of pasture in the cow, the change of milk, over-heated rooms, sudden change of weather, are often beyond control. But as soon as the first symptoms appear, the indications for treatment are absolute. If the digestive derangements are very mild, simply consisting of a furred tongue, eructation of sour milk, or the appearance of curds or slime in the stool without any actual diarrhœa, weakening the bottle with a little more water, taking out the farinaceous material that has probably been added to it, giving a tablespoonful of Murray's fluid magnesia, and allowing the child to drink, if it is thirsty, weak gum-arabic water or toast-water, will probably be all-sufficient. But if the child has pain in addition to these symptoms, the treatment is different. For such cases, stop the bottle-food at once, and when food is required give the child a mixture containing a tablespoonful of cream, a tablespoonful of fresh milk, and two tablespoonfuls of lime-water. Be careful not to overload its stomach; feed it at frequent intervals, very little at a time. If it is thirsty, allow it to drink of the gum-arabic water or toast-water. If the child is a year old or more, when the thirst is very great, and especially if it is in summer, milk can be omitted altogether from the diet, and barley-water—made by adding an ounce of barley, crushed, to a quart of water, boiled for twenty minutes and then strained—can be given with condensed milk, one part to twelve; or to this can be added two tablespoonfuls of lime-water when the bottle is made up, and the child be given every two or three hours ten drops of whiskey, as a drink, in one ounce of water, with one ounce of lime-water. The white of an egg dissolved in a tumblerful of warm water, to which is added a teaspoonful of glycerin and a dessertspoonful of orange-flower water, can be kept in a cool place and given occasionally, as a drink, if the child is thirsty.

In cases just described, the medicinal treatment should be to clear the bowel at once of the accumulated curds and mucus, and relieve the congestion; and for this purpose a dose of castor oil may be given as follows: a teaspoonful of castor oil, ten drops of aromatic syrup of rhubarb, and a dessertspoonful of warm soda-mint (solution). A warm foot-bath should be given, and over the child's abdomen should be placed a mustard

DIARRHŒA. 171

poultice (half flour), a spice poultice, or flannels wrung out in hot water, over which is sprinkled some essence of ginger. After the oil has operated once, and a large mass of curds and mucus has been expelled from the bowels, the subsequent operations will be noted to be of a more natural color, but very cautiously indeed should the mother return to the ordinary bottle-food. If the weather is very warm, the child should be taken daily to some shady spot in the suburbs, where the air is purer and fresher than in the city. If it is at all possible, it should for several hours a day breathe the air that comes off the water. There is no city charity that should receive more encouragement than that which gives the river excursions to these poor children. A few moments' ride in a ferry-boat will often revive a child that is being exhausted by the heat and succumbing to summer-complaint. The nervous system has given out, it has lost its control over the blood-vessels, and the serum or watery portion of the blood, that contains all the nourishment, leaks into the bowel and is drained away. The little sufferer dies absolutely of hemorrhage,—white hemorrhage. That is the reason these teething city babies so rapidly emaciate; that is the reason they crave water to supply that which is lost. A child, then, should be treated in such a way that its nervous system will be strengthened and given once more control. Fresh, pure air,—either the air from the country or, better, the air from the water, or, better still, the air from the sea. Its tissues should be braced by stimulating, cooling applications; it should be sponged with vinegar and water or alcohol and water; its feet should be soaked in water containing a little mustard to stimulate its circulation; the room should be kept as cool as possible, thoroughly aired, free from all contamination by sewer-air or putrefying matters, and what food is given should be absolutely pure and untainted.

In all these cases, if in the city, avoid milk that is raw, and starches of all kinds, even should they be thoroughly cooked or boiled. After a child has taken a dose of oil, it can be fed on condensed or evaporated milk, and wine-whey given to it as a drink, a wineglassful every three or four hours should it exhibit any symptoms of exhaustion ; and after the bowels have thoroughly returned to their normal condition, the food can be changed to either boiled milk prepared for the bottle, or some of the other preparations mentioned in the chapter on that subject. In these cases peptonized milk is often very valuable.

It is well for a mother to recognize the fact that her child should abstain entirely from food in these earlier stages of diarrhœa, especially in summer, certainly until her physician has seen the case. She should make it a rule to allow the child to drink freely, if it wishes it, of gum arabic or toast-water ; give small pieces of cracked ice, wine-whey, Valentine's beef extract or the fresh expressed beef-juice, or some chicken-broth ; indeed, the beef-juice and chicken-broth, given with gum-arabic water, make a nutritious and harmless diet in all of these cases. This hygienic treatment alone, in the earlier stages of a diarrhœa, will be of very great advantage, and will aid the doctor most assuredly by paving the way for that medical treatment which he may find it necessary to institute to bring about a normal condition.

Of course, a diarrhœa may come on very suddenly, with great intensity, and be fatal within a few hours. These cases are called cholera infantum. Their suddenness is appalling ; their treatment should be immediate. Their cause seems to be a union of all those conditions, one of which alone is sufficient to cause a diarrhœa ; it is a blood-poisoning, originating, possibly, in the absorption of some germ of putrefaction, and aggravated by improper diet, lowered vitality on the part of the child, impure air, and intense heat, continuing unabated day and night. Cases may originate in this way, coming suddenly upon children who have been previously in good health, or this condition may be the result of a protracted and neglected simple diarrhœa. The mother's duty in these cases is always to attend to the hygiene, and by hygiene we mean the attending to proper food, judicious bathing, and the placing of her child where it can, for at least a part of the day, breathe pure air.

But the doctor's duties are equally important. My object is not to instruct mothers how to take the doctor's place,—his duties are distinct from hers; but it is simply to give her an intelligent understanding of what the doctor means when he lays down certain rules which are to be carried out in the nursing and management of her child, without which the medical treatment would be absolutely of no avail. It would be ridiculous to give a child morphia and sulphuric acid for a severe case of cholera infantum, or salicylic acid or bismuth and pepsin, and at the same time to have the baby kept in a close, ill-smelling room, and endeavor to quench its thirst by the free administration of tainted cow's milk given through a dirty nursing-bottle; and yet how often, even in families that should know better, do we find appalling ignorance in regard to these most important matters.

If the child vomits much, food by the mouth should be avoided and nutritious injections given instead, such as a raw egg beaten with warm water and five or ten drops of brandy or peptonized milk, about one ounce every two hours.

CHAPTER XV.[1]

WHOOPING-COUGH.

Its Character—Its Complications—The Nursing of it—Diet—Treatment.

It is not whooping-cough alone that gives us such a large mortality: it is fatal in its complications. In other words, being a disease that lasts six weeks, or sometimes two months, parents or nurses get careless and the babies die from debility, pneumonia, bronchitis, bowel-complaint, or some complication caused by the great straining induced by the cough.

Let us for a moment study the nature of this disease from the most recent investigations of the subject. It has been well known and thoroughly described by medical writers for many years; certainly for over three hundred years it has been acknowledged as a contagious affection, not simply a nervous cough.

It is a disease usually occurring in childhood, but which is limited to no particular age. I once saw a child born with it undoubtedly, and we frequently hear of it at the age of sixty or over. It is rare for a person to have it twice. It begins as an ordinary cold, though, after the child has had apparently a cold in its head and a dry cough for some days, the cough comes on in paroxysms, usually upon taking food, or following excitement; but it tries to suppress the cough, and in doing so the face gets much congested. Soon the paroxysm of coughing will provoke vomiting, and then the *whoop*, or drawing-in of the breath,

---

[1] This chapter appeared as an article in *Babyhood* for September, 1886, and is republished by permission of the publishers.

after several forcible expulsive coughing-spells have taken place. These paroxysms increase in force and frequency; the child's face becomes blue and puffy; the little sufferer seizes on the nearest obstacle for support; the spells are agonizing to behold; finally, after successive coughing-spells, each followed by a gasping which produces the whoop, a violent attack of vomiting takes place, mucus runs from the mouth and nose, and the child falls back limp and exhausted, in a few moments to resume its play, without any evidence, except the puffing about the eyes, of any trouble in the respiratory tract. This is the characteristic feature of the disease.

It is impossible at first to make a definite statement as to the existence of the disease; it is only after the first week or ten days, when the cough assumes the features above described, that one is certain; and especially is this the case if the child has been exposed to the disease. It is not the most contagious disease of childhood, not nearly as much so as measles, chicken-pox, or scarlet fever; it requires, in all probability, either the breath direct from another case, or inhaling the air of a crowded room or car in which a child with the disease has been.

In the article on scarlatina the period of incubation is explained,—that time which elapses between the direct exposure to the poison of a disease or the absorption of its contagious principles, and that of the appearance of the first symptoms. For whooping-cough this is supposed to be about two weeks, as nearly as we can calculate it.

The disease is a contagious one—that we all know. Its contagious material resides undoubtedly in the breath of the one affected with it; it is taken up probably by the air during its passage over the secretions from the throat and nose. The air probably retains it for a long time, and these secretions usually, at the height of the disease, being excessive, are easily communicated from one child to another in play, by kissing, etc., or they dry and are finely pulverized, and are scattered throughout a badly-ventilated room. During the whole period of excessive secretion, from the first week till the cough almost disappears, the child with whooping-cough is capable of giving the disease to others with whom it comes in contact.

It is almost criminal, then, for parents and nurses to take these children into open squares or parks, places of amusement where healthy children congregate. And yet this is done daily.

Investigators have detected by the microscope certain germs, which have since been found in all cases of this disease, in the sputum and in the secretions from the nose. In all probability these come in contact with the mucous membrane, are directly absorbed, multiply and diffuse themselves through the blood during the period of incubation, or act in some way especially upon the delicate membrane of the nose and respiratory tract. Animals have been inoculated with this mucus and have exhibited all the symptoms of the disease.

It is a disease that has a regular course of its own to run if not treated. In Japan it is called the "hundred-day disease." It usually, in mild cases, lasts about two months, possibly sometimes three, though active treatment and careful nursing will shorten it to a very great extent, or make it very much less severe.

It is not for us here to examine carefully the medical treatment of this disease. As far as is known, no drug will at once cut short an attack, but the physician has many resources at his command that will mitigate its severity, diminish the intensity of its paroxysms, lessen the great danger to hemorrhages, to heart-disease, to disease of the lungs, whose delicate tissue is easily damaged by forcible coughing; and it is for him to decide what is especially advisable for each particular case. He will probably give the child belladonna or quinine, or possibly recommend an infusion of chestnut-leaves. At night he may find it necessary to give the bromides of potassium or ammonium, with chloral.

To the mother belongs the equally important duty of warding off danger by careful nursing, by guarding her child against exposure, by proper clothing, careful and nutritious diet, and abundance of fresh air.

Every child with whooping-cough should have pure, fresh air, and be out of doors as much as possible in suitable weather. The sea-air, when accessible, is greatly to be desired; the moisture in it, and possibly the salt, help to liquefy the secretions. It has a sedative effect, allaying nervous irritability; then it is always fresh and pure, and this is most important in the treatment. The fresh air prevents debility, and therefore wards off those serious complications due to "run down" and loss of appetite. If the patient is obliged to remain in the city, it is well to dress the child warmly, make a playroom of the top story of the house, and open the windows to insure a constant access of fresh air, purer than that directly from the street.

Diseases of the lungs are most to be dreaded as a complication, therefore all children suffering from whooping-cough should have some woollen garment, light in summer and heavier in winter, covering the entire body. It is a mistake to clothe children too warmly; active perspiration during play will only invite colds should the child be exposed to a draught. Indeed, this is the case with all children, sick or well. Often severe catarrhs are brought about in children overclothed.

A child should be accustomed to a daily bath or sponging in cool water; its chest, back, and *feet* should be thoroughly sponged, and then reaction brought about by a good rubbing.

There is a curious close relationship between the feet

and the mucous membrane of the air-passages. We all know how quickly wet feet, when exposed to draughts, will give a catarrh. A foot-bath of cool or cold water at night will do much to lessen the liability to colds.

A child with whooping-cough should receive a full supply of digestible, nutritious food; gruels, soups, or broths; Mellin's or Horlick's food; a light dinner of chicken or steak, if it is old enough, vegetables such as well-cooked young beets, spinach, cauliflower, roasted or mashed potatoes, and stewed fruits, as apples or peaches, that will keep the bowels regular. Milk should, of course, be plentifully used, and it is well to dilute it with lime-water, or Vichy water, to prevent it curdling in heavy masses. A child is apt to have a paroxysm and vomit a meal just after it is taken. Some light, nourishing food should at once be given which will be retained,—either a cup of broth with barley, or milk and lime-water, or some sherry and water with sponge-cake, will serve the purpose.

I have often noticed that a copious discharge of a thick mucus will follow a paroxysm, and that until this discharge takes place the cough will be repeated. The mother should recognize this fact and do all in her power to aid the child. If it is an infant it can be accomplished by turning the child over on its stomach, with the head low, and detaching the mucus with her finger from the mouth. If the child is older it should be taught to blow its nose thoroughly, or probably by inducing vomiting at once relief will come and the paroxysm be cut short. A child can be very materially aided in this way and the severe strain avoided.

A large amount of the thick mucus which cannot be gotten rid of by the cough may seriously threaten suffocation. In such a case a teaspoonful of syrup of ipecac, or the same of powdered alum mixed with syrup, should be at once given to provoke immediate vomiting.

Counter-irritation to the chest by means of various

liniments is certainly valuable; a suitable one will be selected by the physician, containing possibly the oil of amber, croton oil, or turpentine, in soap liniment, or possibly chloroform liniment. My own experience has taught me that the spray from an atomizer is a valuable aid in the treatment of this disease. It can be used to make the thick mucus more watery, and also as a means of carrying medicaments directly to the surface. But it requires an immense amount of patience on the part of the mother. The nose and throat should receive in this way a thorough treatment five or six times a day. The nose should be frequently cleansed by the spray of Dobell's solution (carbolic acid, borax, soda and glycerin).

I have endeavored to impress upon my readers that whooping-cough is a distinct disease with a history of its own, and belongs to the same category as the other contagious affections of childhood. I have avoided discussion of its medical treatment; that belongs to the doctor. Whatever tends to weaken the child invites most serious complications, and these are the causes of the great mortality mentioned. All patent nostrums of uncertain combination, or of such composition as to derange the secretions, produce constipation, loss of appetite, are to be avoided by all means. Good nursing is more than half the battle. Fresh air, proper clothing, good, nutritious diet, regulation of the bowels, efforts to mitigate the intensity of the paroxysms by aiding the child in discharging the mucus, giving it an easy and comfortable position which relieves the strain, supporting it for the same purpose, pressure of the hand over the eyes to prevent their strain,—all these are as important as the medical treatment.

CHAPTER XVI.[1]

SCARLET FEVER.

Its Cause—The Reasons why it does not appear to be as Contagious as other Similar Affections—The First Symptoms—Its Nursing—Complications—Sequelæ.

SCARLATINA is the Latin name for scarlet fever in all its forms, those most grave and those most mild. There is another point of great importance, which is, that the mildest cases of scarlet fever have often the saddest ending, owing to the want of attention to those details of nursing which exclude the possibility of serious complications by exposure or neglect. Every doctor has heard the statement, when questioning in regard to the antecedents of dropsy or possibly some destructive disease of the ear: "My child had only scarlatina; it was so mild that really we did not see the necessity of punishing the little one by confining her to her bed or room."

Scarlatina is scarlet fever, and a mild attack of this most treacherous disease may become as serious ultimately as a very severe attack,—often more so, as the one will be cared for and the other neglected. It is, therefore, necessary thoroughly to understand the principal features of this disease.

It is undoubtedly caused by the entrance into the system of a *something* which has all the characteristics of matter. This *something*, which, as we shall see, requires a certain time thoroughly to infect the blood in

---

[1] This chapter appeared as an article in *Babyhood* for February, 1886; it is now out of print, and is republished by permission of the publishers.

which it circulates before the symptoms of the disease show themselves, is called a *germ;* this germ has *weight,* proven by the fact that it occupies a certain space ; can be carried in clothing or merchandise, or by the air ; it remains in its position ; it does not evaporate or become gaseous ; it is transmissible, and at the same time is very subtle, can insinuate itself in cracks and crevices, in the breath of individuals, in their hair, in clothing, in the nails. It is a living material, whose vitality may lie dormant for years, and then, like grain, grow under favorable conditions ; it can be destroyed by heat, probably by intense cold, or by chemical agents. But it is unlike the grain in one most important characteristic, which is this : a grain—say of wheat—may have remained dormant since the time of the Pharaohs, and, taken from a mummy-coffin, placed in heat and moisture, it will sprout, but will only produce *one* stalk. The germ of scarlatina may lie dormant, but when aroused into activity by suitable associations it will act as a leaven—as a ferment ; this minutest microscopic object will perpetuate its species until it will invade every organ and tissue of the body. On this account this disease is classed as *zymotic* (from *zyma,* ferment).

Scarlet fever can be communicated by infected milk ; and, as far as we know, the milk has only to stand in the room where the disease exists or has existed, to absorb the germs, which are so subtle, so light, and yet so tenacious, as to float in the air and adhere to particles of dust.

We all know how much dust is constantly floating in the air ; let a beam of sunlight pass through an opening in the shutter, and we can readily see how the scales of the skin from the body, pieces of lint, etc., can carry these microbes which may be thrown off in the mucus from the nostrils and mouth or in the perspiration, and even in the urine.

Not only are these secretions **germ-carriers,** that is,

contagious,—and they have all been proven so by direct inoculation,—but the passages from the bowels as well as the urine are so. In that way sewer-air may be a means of their conveyance; drinking-water also, as well as the vapor from the soil on which these matters have been thrown. Bear in mind, then, that the scarlatina poison can be carried in this way hundreds of miles; that it does not need the personal contact of individuals; that it retains its vitality for months and even years unless it be subjected to certain influences that either entirely destroy, or deprive it of, its malignancy; these are intense heat, especially boiling or steam, plenty of fresh air, and certain chemical substances, as chlorine, sulphurous acid, and others.

The poison of scarlatina is either inhaled by the individual, or swallowed; it is then taken up by the circulation, and, finding itself surrounded by material which develops it, vivifies it, becomes rapidly reproduced, and the symptoms of the disease show themselves. This period between the reception of the poison and the appearance of the symptoms is called the period of incubation; this is known to be from one to six days; in some cases longer.

Unless we definitely know that the child has been exposed to the disease, we cannot detect its presence till the rash appears, and this takes place within twenty-four hours of the first symptoms. What are these? Chill or convulsions, delirium, intense headache, sore throat, swelling of the glands of the neck behind the jaw (kernels), nausea or vomiting, associated with high fever, bounding pulse, and dry skin. The first three of these may be absent in mild cases; the others are nearly always present to a greater or less degree.

Under such circumstances what should be done? The child should be put to bed by itself in a separate room; it should have a hot foot-bath, the water, with or without mustard, about as warm as the hand can bear. It should be lightly covered with a sheet and light blanket; it should be kept as quiet as possible, and given frequently warm milk in small quantities, with lime-water or so-called "cambric tea," and no solid food. Do not purge; in fact, avoid all medicine —with the exception possibly of a little sweet spirits of nitre, a teaspoonful to half a tumbler of sweetened water—until the doctor arrives. At the end of from six to eighteen hours the rash will appear. It will be noticed in patches, fading into the normal color of the skin, on the neck, shoulder, and chest, then on the abdomen, and finally on the trunk, arms, and legs. It resembles the redness produced by a mustard-plaster, and feels rough to the hand when fully developed.

The disease is now fully determined; the fever is high, the restlessness is usually increased, the throat symptoms are marked, and the secretions are diminished. The most urgent care is now to be taken in the nursing. All superfluous hangings, such as curtains, pictures, should be removed from the room. The child should be nursed by one who has all the details of the case under her charge, who should wear the simplest kind of clothing, that can be daily changed and washed or aired. She should have an adjoining room in which to keep her clothes and make her toilet. Everything that comes in contact with the child, such as towels, brushes, blankets, or sheets, should be kept rigidly separate, and thoroughly boiled and aired before being taken from the premises.

The room should be kept thoroughly ventilated, either by keeping open a window in the adjoining room or by some arrangement attached to the window of the sick-room which will allow the ingress and egress of air without a draught; the temperature should be kept at about 68°, and regulated by a thermometer. If the room receives its heat from a furnace, the hot air should be made to pass over a pail of water containing either diluted Labarraque's solution or Platt's chlorides, and a towel with one end dipped in such a solution should be tacked over the register. If there be a stove, or, better than all, an open grate, these solutions can be placed near by, so as to be readily evaporated and distributed throughout the room.

The chamber should always contain some such solution in which to receive the excreta. A small quantity of urine should daily be collected in a clean vessel for the doctor's examination. It is usual to anoint the child with some greasy substance; this allays the intense itching or prickling, which is most annoying, it softens the skin, which is inflamed and swollen, it depresses the fever to a certain extent, and it serves to collect the scales of the skin, which, if shed, serve as carriers of contagion, and which are usually shed in flakes. The child should have its mouth washed once or twice daily, as also other parts of its body, for purposes of cleanliness, and the water used can contain either very diluted Labarraque's solution, vinegar, Listerine, or phenolsodique, and possibly the doctor will order the frequent use of the hand-spray, using some good disinfectant for the throat in these cases.

What are the dangers incident to scarlet fever? Extensive disease of the throat with complications of diphtheria, disease of the ear with permanent deafness, disease of the eyes; more important than all, serious complications due to inflammation of the kidneys, made evident by dropsy, convulsions, often ending fatally. To avoid these, which may take place in the mildest cases, from exposure to draughts, imprudence, and want of cleanliness and attention, great care is necessary.

The question of bathing or sponging a child ill with scarlatina must be decided by the doctor. Sometimes it is necessary to depress the temperature, as a prolonged high temperature will kill; but in all cases, however severe, cleanliness should be insisted upon,—the face

SCARLET FEVER.

and hands, the eyes, ears, mouth, and genitals, should be kept clean and free from secretions.

The temperature usually remains high till the decline of the disease,—about the fourth or fifth day in ordinary cases. As soon as the fever has subsided and the eruption has faded, and the skin-shedding is well established, it is customary to sponge the body off thoroughly in tepid water, and clean the head, using a fine sponge or soft linen, avoiding draughts, and keeping the body well covered, with the exception of the part being washed. I have found a preparation known as "Little's Soluble Phenyl" admirable in this connection, a few drops of it being added to the water. It is disinfectant and leaves the skin soft. As kidney-troubles usually show themselves during or following the scaling stage, greater precautions than ever are to be used at this time. The urine should be examined every day or two. The diet should be mostly liquid,—that is, milk, or milk and lime-water, gruels, soups, and such like; the child should be encouraged to drink freely of water, the bowels must move daily, if necessary by an enema, and under no circumstances should the patient be permitted to leave the room unless great precautions have been previously taken. In this climate we have to be very particular, owing to the sudden changes of temperature, and it is far better that the child should be kept in-doors a few days longer than the parents usually think necessary than to run the great risk of kidney-diseases, or rheumatism with its serious effect upon the heart, which may follow such exposure.

After the child has had several changes of underclothing, has been well washed a number of times, and at least two weeks have elapsed since the disease declined, it can be removed to another room, and the sick-room fumigated. This should be done by igniting some sulphur in a saucer in the room, all the windows and doors having been previously closed and the cracks stuffed. After twenty-four hours the room can be opened and full ventilation permitted. All the furniture should be wiped with a damp cloth, and the paint-work washed with water containing the chlorides or borax. The room should remain unoccupied for some time and be thoroughly aired.

The school-room is undoubtedly the place most to be blamed for the distribution of scarlatina poison. To get rid of the other children they are sent there whilst the mother is nursing the sick one at home. Some children possess a remarkable immunity from this disease, nevertheless they act as carriers of contagion. Then, again, servants or child-nurses often carry it in their heavy shawls from house to house, taking it directly from a sick-room to the nursery.

Scarlatina, as far as we know at the present time, only comes from previous cases of the disease. *Cleanliness not only lessens the danger of serious complications which are often fatal, and mitigates the severity of an attack, but it is the great germ-destroyer, and prevents the spread of this dread disease in households.*

CHAPTER XVII.

MEASLES.

How it is contracted—How the Contagion is carried—Why it is the most contagious of the Eruptive Diseases—The Peculiarity of the Eruption—The Dangers of Pulmonary Troubles as Complications or following the Disease—The Nursing.

THERE are so many matters in common between scarlet fever and measles, that it will be unnecessary to repeat. Measles is a disease which comes from a special poison of its own, one which only produces measles; this poisonous principle, or germ, is especially active in the breath, in the secretions from the eyes and nose, and the whole respiratory tract; and as these secretions are very much more active—in fact constitute a marked feature—in this disease, the child with it forms a focus of contagion for those who come in contact with the air which has passed over the surface of its mucous membrane. The anointed skin in scarlet fever, the absence of marked catarrh and profuse secretion, the thorough isolation which is always insisted upon, probably account for the fact that measles seems by far the more contagious disease. There is another peculiarity which also accounts for this: the catarrh of measles, resembling an ordinary cold in the head, may be mistaken for such, and the child for the few days preceding the rash may associate with other children, and thus disseminate the elements of contagion. The period of incubation in this disease is from twelve days to two weeks, but of course during this time, unless we know a history of exposure, there is nothing special to attract our attention. The disease usually manifests itself by all the symptoms of a violent cold in the head: the eyes become suffused, very watery, and intolerant of light; the discharge from the nose is constant, the child's face puffy and red, apparently swollen; there is a tendency to drowsiness. Of course there is fever, the child's skin is hot and dry, and the little sufferer rolls and tosses from side to side. These symptoms are not very marked at first, and the child is supposed to have caught cold; but they increase in severity, and it is not until about the fourth day when the rash makes its appearance upon the face. Previous to that, however, if the throat be examined, the outline of a rash characteristic of measles can be determined upon it. All this time the child suffers with paroxysms of a dry, ringing, croupy cough; the tongue is usually slightly coated, the appetite is lost, and the fever, with the catarrh of the mucous membrane, gives rise to intense thirst. The first appearance of the rash is upon the temple, the forehead, the neck, extending down the chest and arms, and finally covering the body. The word rash is a misnomer when applied to the eruption of measles, and refers more especially to that in scarlet fever. When noticed upon the forehead, the temple, or the neck, it seems at first as if it were beneath the skin; a number of small clusters of points, resembling flea-bites, that form a crescent when the finger is passed over the surface; a slight elevation is noticed. These elevations increase until they rise distinctly above the surface, and form a papule. As soon as the whole surface of the body becomes covered with the eruption, the skin is decidedly rough and papular to the touch, and the crescentic outline of the papules in these patches over the entire surface of the body can be distinctly noted. There is no mistaking a case of measles at this stage, or confounding it with scarlet fever. The suffused eyes intolerant of light, with swollen eyelids, the puffy face covered with its speckled eruption, the excessive nasal discharge or evidences of swelling of the mucous membrane of the nose, the croupy cough, which is constant and annoying, are in themselves sufficiently plain to be the distinguishing feature. The rash after three or four days gradually fades from the surface, leaving in many cases a slight staining of the skin in freckles or spots, disappearing first from the parts first affected. The fever gradually subsides, though the cough and evidences of bronchitis may remain for some time longer. The skin will not come off in flakes, as is usual in cases of scarlet fever, especially where no ointment has been used; it is usually shed in small bran-like scales.

The lungs bear the brunt of this disease in neglected or severe cases, and just as we guard the kidneys in scarlet fever, to prevent their congestion, so in this disease we protect the lungs from the very first, by carefully guarding against draughts, impressions of cold, internal congestion, by endeavoring to establish the eruption over the whole surface of the body, by protecting the eyes from light and the skin from draughts. Measles is a disease which is very fatal among the poor, much more so than scarlet fever; that is, in the earlier stages, owing to the fact that acute diseases of the lungs, as congestion, pneumonia, kill very much more quickly than the diseases of the heart and kidneys that follow scarlet fever, and probably do not show themselves for weeks or months after the termination of the disease. Croup in all its forms, bronchitis, congestion of the lungs, pneumonia, may occur in the earlier stages of measles and be fatal, through exposures to draughts and through improper nursing; nor, indeed, is a child entirely safe until all evidences of catarrh of the respiratory organs have disappeared. Taking a child into the cold air too soon may bring on a fatal pneumonia; exposure to sewer-gas in the sleeping-room may produce a fatal diphtheritic croup; the exposure to the contagion of whooping-cough may engraft this disease, with serious results; the throwing off the bedclothes, thus chilling the body, may congest the lung or eventually affect the heart; then, again, indigestion may produce an irritation of the bowel, and a serious diarrhœa follow. The mucous membrane of the stomach and bowel is affected in this disease, simultaneously with the mucous membrane of the respiratory passages. Although vomiting is not as apt to occur as an initial symptom as in scarlet fever, still diarrhœa, especially in the heated term, is often a serious complication, probably indicating inflammation of the bowel, and should be carefully watched. Not alone do we have complications of so serious a nature as to threaten life, but in scrofulous children, or those who are simply run down, loss of sight or of hearing may be the result of careless nursing.

The same general principles regarding the ventilation

of the sick-room, cleanliness of the body, diet, ought to be observed in the nursing of measles as we have just laid down for that of scarlet fever, possibly slightly modified, owing to the difference in the two affections. Thus, in measles the tendency to pulmonary congestion will necessitate keeping the feet warm to promote circulation, and possibly the use of poultices to the chest, or of cotton or some non-conducting substance. A child with measles should be guarded as carefully against going out too soon as in scarlet fever. I may once more impress upon mothers the fact that if they have a dry, well-ventilated, sunny nursery, apart from the sleeping-room, it is a mistake to run any risk in taking a child out of doors when the weather is the least objectionable,—that is, on days that are damp, raw, and foggy, or when the winds are piercing. I insist upon this, not only for children who are convalescing from disease, but even for children who are perfectly well; there would be fewer catarrhs and sore throats if this plan were more generally adopted. Great care should be taken, when a sick child has a movement from its bowels or bladder, that it is well covered and protected from draughts.

I need not repeat all that has been said when speaking of scarlet fever, in regard to the diet and home treatment; the mildest form of nourishment should be given,—milk, diluted with an alkali, given at frequent intervals, also chicken-tea, egg albumen in water, beef essence or juice, with cracked ice, toast-water and wine-whey, or wine-whey and barley-water. No solid diet should ever be given in fevers. The child should be allowed to drink water freely, but in small quantities at a time. Glycerin and water, the proportion being a teaspoonful in half a tumbler of water, will often relieve the dryness of the mouth and throat, at the same time allaying thirst. Weak lemonade sweetened with a little glycerin, not immediately after milk, is most refreshing in fevers.

Sweet spirits of nitre, a teaspoonful to a tumbler of water, will quiet the nervous system, if the child drinks frequently of it during the night, and is likely to promote sleep. A hot foot-bath is always efficacious, if the feet are kept warm afterwards. A teaspoonful of spirits of Mindererus, with a teaspoonful of sweet spirits of nitre in a half-tumbler of water, forms a household fever-mixture which can be safely sipped by a child from time to time until the doctor arrives. Anointing the body as in scarlet fever, with carbolated vaseline or cold cream, allays irritability even from the onset of the eruption. The room should be darkened, as the eyes are weak.

CHAPTER XVIII.
SECOND DENTITION.

Forcing in Education to be Condemned—What is Meant by Second Dentition—Complications Due to Hereditary or Acquired Conditions—Rickets—How it is to be Avoided—Diet of Children at this Age—Clothing.

THIS work would be incomplete were we simply to devote our attention to that time which is limited to the cutting of the milk-teeth. There are various disorders of childhood that are dependent more or less upon the disturbance of the equilibrium that should be maintained between growth and development. This disturbance is usually brought about by the unnatural pressure which fashion or habit exerts upon the growing tissues of the child, by brain-forcing and muscle-cramping, by the want of that freedom and abandonment which gives food for the muscles' growth and carries away the ashes of their destruction, and the pernicious system of over-education, that endeavors to place an adult brain in a child's body. Indeed, it is this attempt at forcing that is productive of the many disorders which attend the period of permanent dentition.

It is not the cultivation of the mind that physicians object to, but it is the system by which it is often accomplished that is radically wrong.

After a child has passed its first dentition, there is little to call our attention, barring the accidents that may arise from contagious diseases. About the sixth year it will cut the four first molars; about its twelfth year it will cut the four second molars; and after the eighteenth year the so-called wisdom-teeth. These teeth are all new ones. The jaw has to change in shape and size to accommodate them as they grow; they produce a certain amount of pressure upon the nerve-pulp at their base, and give rise to neuralgia and reflex disturbances, produce irritation of the mouth, the throat, consequent disorders of the tonsils, ear-troubles, and, possibly, disorders of the intestinal canal, interference with digestion, with blood-making, and, possibly, as a consequence, many of those peculiar hysterical phenomena that attend puberty, both in boys and girls, but especially girls. In fact, a child's brain that is excited by over-study, over-stimulated, in other words congested, and by blood that is not of the purest, owing to deficient exercise and under-feeding, falls a ready victim to the slightest pressure or irritation that will result from the growth of one of these permanent teeth. The other sets of permanent teeth, that are cut from the sixth to the fourteenth year, simply take the place of milk-teeth, and, consequently, do not produce the same degree of irritation; it is, indeed, a difficult matter to decide in an individual case exactly how much disturbance is due to teething, and how much to the general impairment of nutrition which results from the child's surroundings and mode of life.

A strong, healthy child, brought up in the country, rarely has any difficulty at this time.

The general impairment of nutrition, which is followed by debility, languor, all resulting from the loss of, or perverted, appetite, and defective secretions, are the most frequent disturbances that we meet with during second dentition. We have at this time, also, disturbance of the mucous membrane of the mouth, of the gums, due to the irritation of the milk-teeth, whose roots become absorbed; possibly, want of cleanliness may lead to incrustations or spongy condition of the gum, with pressure and ulceration. Also enlargement of the tonsils, with various forms of inflammation, are very apt to occur at this time; especially that variety which is attended by a grayish deposit, very readily mistaken for diphtheria, and to which the physician's attention should be immediately called. We also have a tendency to chronic enlargement of the tonsil, which will give rise to earache, sometimes difficulty in swallowing, snoring at night, with restlessness, and a tendency to acute inflammation or quinsy upon the least exposure to cold.

Such children are prone to have attacks of neuralgia, headache, toothache, or neuralgia of the face; their tongues are apt to be coated; they have frequent bilious attacks. It is also at this time that what is known as scrofula is apt to develop, enlargement of the glands of the neck, weak eyes, nasal catarrh show themselves; all these conditions, although occurring at the same time as the second dentition, are by no means dependent upon it in all cases; they are attendant upon excessive growth and faulty development, malnutrition, inherited taints. During this time also occur frequently those deformities which are the result of improper deposit of lime in the bone, or the reabsorption of bone from pressure due to all sorts of mechanical causes; among these we find the diseases of the joints, curvatures of the spine, and also the imperfect formation of the permanent teeth, which renders them brittle or soft, readily broken, or easily affected by the acid eructations that come from a disordered stomach, —a mere shell of a tooth, as it were, which soon becomes carious, and unfortunately cannot be replaced.

We see, then, that the mother's attention should be called to her rapidly-growing child, in order to perfect its development in such a way that its tissues will be well formed and nourished, its functions established, by the time it reaches maturity. There are certain other disturbances which occur during second dentition that require our attention; there are frequent bilious attacks, or certain fevers apparently due to growth, that come on without any appreciable cause. A child will have a slightly furred tongue, will complain of pain in its bones, of languor, loss of appetite, will be restless in its sleep, will suffer from headache, which is usually frontal. Suddenly it will develop a high fever, the face will be flushed, the eyes red, and indeed if vomiting occurs, and the child has not previously had the disease, we might almost expect the onset of scarlet fever. The child should have a hot mustard foot-bath, should be given sweet spirits of nitre in its water to drink; if the tongue is coated, it should be given a dose of either Murray's or Phillips's fluid magnesia, or a dose of castor oil, and the chances are the next morning the child will awake, bright and cheerful, without any bad results from its attack.

Rheumatism is very common at this period of life; indeed, many of the cases of so-called growing-pains are nothing more or less than masked rheumatism. If a child complains constantly of aching in its limbs, seems loth to take exercise, the attention of the doctor should be called to its condition.

Elsewhere I have written of the care of infants during the period which terminates at about the age of five years. There is not much to call our attention before this time to any difference between the sexes; the girls are usually more delicate, the boys hardier; the girls having more tender skins, more liable to eruptions. Boys are more difficult to bring up, owing to the fierceness with which they are attacked by infantile complaints. Usually the difference in sex makes itself manifest in temperament as soon as the infant becomes sufficiently independent to amuse itself, and especially is this the case with the first-born, who has no one to copy, no playmate older than itself by whom it can be led. Even at this early age parents should as much as possible endeavor to train a child to enjoy out-door amusements, to give it exercise, that it may grow hardy, graceful, and cheerful in disposition. This will have to be encouraged in the girls; the boys take to it as ducks take to water.

It is in these tender years, when the bones are yet frail and soft, that those deformities which become the source of so much disease in later life develop themselves. People are apt to imagine that rickets and scrofula are confined chiefly to the lower classes, to the poor. This is a great mistake; they enter every household where the conditions which favor their development are found.

Let us dwell for a few moments upon this matter of importance in infant life. All the long bones of the body are composed of three parts; the two ends which come in contact with the same part of other bones, and form the joints and then the shaft. In later years these ends become united to the shaft, and the whole becomes one solid bone. Bones are composed of two kinds of material: the gelatinous, which gives pliability and elasticity to them, and the mineral, which makes them firm and strong. Of the latter, the salts of lime form the most important part.

In infant life the growing bone receives an immense amount of nourishment; it grows rapidly. The centre is spongy, and filled with blood-channels. The bones are by no means hard at this age, and should anything diminish the amount of lime they receive, the movements of the child, the straining of its muscles, the weight of its body, would cause the bones to curve, and unless attention is immediately given, this curving will be permanent. Especially is this the case where the bony shaft is attached to the cartilage to form a joint; naturally, the rapid growth requires such joints to be loose and yielding and elastic, and the constant strain will so stretch them as to produce deformity.

A young child who, instead of running out in the fields, or gayly skipping up the street, filling its lungs to the utmost and sending ample nourishment to every tissue of its body, sits by the hour by its mother's side nursing her doll, though an object of beauty is a melancholy one, as time will show that nature's laws are being interfered with. We know that there is a pressure of fifteen pounds to every square inch of surface on all bodies at the level of the sea. In breathing, a certain amount of force is required to overcome this. In a frail, delicate child, whose muscles are imperfectly developed, the lungs are scarcely ever inflated to their fullest extent, the constant atmospheric pressure not overcome will tend to flatten the chest-walls, and when the bones are hardened the breast-bones assume the form known as pigeon-breasted; one of the great predispositions to consumption will be introduced.

A child that is made constantly to breathe fully and deeply by out-door exercise, and at the same time has its chest freed from all restraints of close-fitting garments or the injurious requirements of fashion, will bless its parents in the future for the great gift of health, which is an heirloom of priceless value. The girls as well as the boys should be straight-limbed and full-chested.

The disease called rachitis or rickets is brought about not only in those children whose parents live in damp, ill-ventilated dwellings, devoid of all the com-

SECOND DENTITION. 197

forts of life, exposed to the intense cold of winter and the great heats and atmospheric impurities of summer; it is also noticed among the rich, who, possessing every comfort in life, may either through ignorance or carelessness fail to supply food which nature has intended to make bones firm and muscles strong, or they neglect to see that their children are kept in a proper state of health to digest and assimilate such food.

Fashionable mothers are only too desirous of transferring their duties to an experienced nurse, and consider the task which was intended by heaven as a sacred duty complete when they have purchased the most fashionable baby-carriage and a full supply of condensed milk. Experiment has shown that the strongest animals can be made rickety by the character of the food with which they are fed, and experience shows that children who show a tendency to rickets, if fed upon nutritious food and given plenty of fresh air and exercise, can be made strong, and outgrow the predisposition. It is on this account that so much stress is laid upon the use of the various cereals in the feeding of growing children, especially oatmeal and cracked wheat. Every child at the age I am now speaking of, that is after the completion of its first dentition, should be given its three regular meals a day: the breakfast consisting of bread and butter, a plate of well-cooked oatmeal with cream and plenty of milk, a soft-boiled egg, a mutton-chop, or a piece of steak finely cut up. Avoid hot cakes, very fresh bread, or rich dishes of any sort. For the dinner, which should invariably be at an early hour, it is an important matter when possible to give fresh vegetables,—well-cooked potatoes and rice, properly boiled, and encourage them to eat bread with their meals. They should also have either cooked fruits or ripe, raw fruits in abundance. The final meal should be varied; meat should be avoided; soft custards, milk toast, or bread and butter made more inviting by a little good preserve. An outline of this sort will give an idea of the character of food that supplies brain, muscle, and nerve. It may be modified according to the peculiarities of the child, the time of the year, and the family circumstances.

Do not let a mother suppose that because she has nursed her infant the usual time it cannot become rickety. Unless she lives upon food which will give to her milk in full proportion the quantities represented in the diet mentioned above, her child, though nursed and to all outward appearances subjected to only what would be best for it, will in reality be no better off than a bottle-fed baby; indeed, a child carefully brought up on the bottle by one experienced should be healthy in all respects.

I think that most mothers forget this important point, that the nourishing quality of their milk depends on their own state of health.

As far as the medical treatment of rickets is concerned, we may only mention here those drugs which are in reality other forms of food. Cod-liver oil is pre-eminently the most important. It may have been ordered by the physician in early infant life, but if at the period of childhood of which we are now speaking, no physician is at hand, a mother should not hesitate to give it should the child complain of fatigue, grow thin, lose its appetite, or show other evidences of bad nutrition.

Before concluding this chapter it may be well to make a few remarks regarding the dress of young children. Habits of regularity should be instilled from the very earliest opportunity, and may be so well established that by the second year the diaper can be exchanged for drawers. Should summer make it convenient to do so at an earlier period, it may be well to make the change, as in some children the thick folding of the diaper is very apt to make them bow-legged, especially if it is not very carefully put on. This of course only refers to children who are walking. As regards the clothing of children in our damp seaboard climate, flannel should be insisted upon all the year round.

For teething children fine merino stockings should be worn all the year. The important things to bear in mind are the simplicity and loose fitting of all their clothing; the latter is most essential for growing children, as deformity, stooping shoulders, contracted chest, and weakened backs depend a great deal upon ill-fitting garments. Would that mothers were sufficiently impressed with this fact!

The next thing is the proper change of clothing to suit the seasons. Warm clothing is not necessarily cumbersome, and merino undershirts with high necks are certainly important even in summer, though they be of the thinnest kind. I do not think that croup is nearly as prevalent as it used to be, and I can but attribute this to the much more common-sense way of clothing children than was fashionable some years ago; the bare-necked, bare-armed, and bare-legged babies in mid-winter are no longer seen,—the frail flowers of our hot winter houses, to be the victims of draughts which are so fatal.

All children should wear thick-soled shoes, easy and well fitting, with room for growth at the toes; they should not have heels.

The importance of light calisthenics at this age cannot be over-estimated. Not only do they give, under competent management, a graceful and easy carriage, which means symmetrical development, but they give muscular power, increase the force of the circulation, develop the chest and thereby the activity of the lungs, give tone to the nervous system, and in other words improve the digestion and the appetite, and in that way secure a flow of good, nourishing blood.

CHAPTER XIX.

PUBERTY.

Puberty—The Four Second Molars—The tendency to Disturbance of Digestion at this Time—The Importance of Proper Food and Clothing, also Freedom from Excitement and the use of Iron—First Menstruation—Menstrual Irregularities—Hysteria—Dysmenorrhœa—The Abuse of Anodynes—The Mother the proper Confidant of her Daughter.

WE will devote a few moments to the consideration of the changes which take place in girls at puberty: partly from the fact that at this time we are still within the period of the second or permanent dentition. Indeed, it will be noticed that at the very age when menstruation first shows itself, the girl is susceptible to all those functional disturbances that may be brought about by the cutting of the four second molars, a set of teeth that are developed anew, not replacing any of the milk-teeth. No wonder, then, that it is most important that the young girl should be under the care of a mother properly instructed to guide her and guard her during this time. The natural impulse of her sex is towards sedentary occupation, seclusion, long dresses, and possibly her first real novel,—all of these in themselves pernicious; they weaken her muscles, lessen her appetite, tend to constipation, excite her brain.

It is scarcely necessary to mention the boys at this age. The out-door life that they lead, especially at the present time, when there is so much for a boy to do, tends to keep them in good physical condition, and renders them insusceptible to the many reflex disturbances which might exist. I may quote from a paper that I recently wrote upon this subject:[1]

We all recognize the very great importance of all

---

[1] Published in the *Medical and Surgical Reporter*, Oct. 23, 1886

that tends towards muscular development in growing girls. They should be symmetrically developed, should have full chests, straight backs, and strong limbs. We should also urge the importance of clothing of light weight and loose fitting, the principal strain being on the shoulders, not on the waist and hips, and also the evil results of cramped, stooped positions in the school-room, eye-strain, and bad ventilation. We all urge these matters daily, and we know how little attention is paid to them. But there are certain forms of various disorders which occur about the time of the second dentition which deserve more than a passing notice. These are manifested either as a chorea (St. Vitus' Dance), nervous excitement such as night terrors, and various mental disturbances (misnamed hysteria), gastro-intestinal disorders, and evidences of malnutrition. The child will probably become languid, suffer with frontal headache, become peculiar in her disposition and show fits of temper, shun society of other children, lose her appetite, become despondent, and possibly develop a local twitching of some of the facial muscles. It is customary to say that this is all reflex,[1] is possibly the warning that the system is undergoing some change preparatory to the menstrual functions,—that it is, in fact, a true hysteria. This may or may not be the case. My own impression is that it is often due to the anæmia (impoverished blood) brought about by rapid growth and development, with faulty assimilation and deficient oxygenation. In my experience such cases present two types, the one essentially nervous, just described, the other the so-called

---

[1] Sufficient attention has not been called to the disturbances caused by the pressure of the twelfth-year molars. These may show themselves in either dental neuralgia, or, in fact, any form of trifacial neuralgia, gastric disorders, or mental peculiarities, amounting to melancholia or symptoms of acute meningeal irritation.

strumous or lymphatic, in whom the want of proper assimilation is shown by a large amount of stored fat, and the anæmia by excessive pallor (called chlorosis).

The nervous system seems to run riot, but this very excitement in itself is an evidence of the demand on the part of nature for a blood-supply which is nutritious and well oxygenated. All the exercise in the world, all the most nutritious and sustaining of foods, will have no effect, until the digestive organs are made to perform their normal functions. If you examine the tongue you will find it coated; the breath is heavy, the bowels are sluggish, the appetite is perverted, the child craves extraordinary articles of food, especially acids and sweets. She has a disgust for her regular meals. There is flatulence, cardiac palpitations, often asthma after exertion. The urine is either scanty and high-colored, or very copious and light. If the menses have been established they are scanty, colorless, and irregular, or there is a leucorrhœa. In these cases the recommendations of popular writers for gymnastics, friction, milk diet, etc., are admirable after the digestive organs have been cleared of their accumulation, and the normal functions whipped into activity.

As far as the general treatment is concerned, the patient should be sponged every morning with tepid water; she should stand in a tub, and have a pitcher of it poured down her spine from the nape of her neck, and then be thoroughly rubbed with a soft Turkish towel into a glow. The breakfast should consist of warm milk, or cocoa, a soft-boiled egg, a rare steak or chop, either oatmeal, cracked wheat, grits, or Indian meal alternating; bread and butter, *not hot cakes*. For dinner, soup, rare meat, fresh vegetables, ripe or stewed fruits. For supper, stewed fruits, bread and butter, warm milk or cocoa, neither tea nor coffee. She should retire early, and not be permitted to read at night. The supply of oxygen should come from out-door exercise, not an over-indulgence in walks or games that fatigue; let the school hours be limited to the early part of the day, and avoid that abomination of preparing lessons in the afternoon or evening for next day's recitations.

Moderate calisthenics, or massage, should be daily given. In about a week's time the girl will be able to bear iron alone, and the tincture of the chloride can be given in ten- or fifteen-drop doses for some time, or a chalybeate water can be given with arsenic. The digestive organs will now also tolerate milk in large quantities, provided it is of medium richness, is fresh, and given warm.

But this is not all. There are very many cases of a highly nervous type which, despite the most careful treatment, will not improve at home. The constant association with parents of like temperament, however solicitous they may be in carrying out instructions, is of itself a cause of nervous irritation.

It may be necessary to send such children from home, either to some relative, living possibly in the country or some distant city, or perhaps to some suburban or country boarding-school, where a thorough change of air and scene, the association with girls of a different temperament, will work wonders.

For the strumous type, the same preparatory treatment may be instituted, and for such I would not hesitate to push the iron, phosphates, cod-liver oil as soon as possible. Change of air to the sea-shore is advisable. There is little trouble in the home-treatment of these latter cases; there is rarely a conflict of authority in such families.

The age at which menstruation appears cannot of course be definitely stated, so much depends upon climate, race, social position, and family peculiarities. Ordinarily, about the age of fourteen may be taken as the average in the temperate climate. Of course whatever tends to early development, such as warm climate, in-door occupation, and especially among those whose occupation is sedentary or where it is attended with much mental excitement pertaining to literary pursuits or the excitement from the whirl of society life at too early an age, will bring about an early appearance; whereas the contrary will have a retarding effect.

There are cases on record which are remarkable for the early age at which menstruation was established; one I may quote from Reinvillier, who records a case reported by Dr. Beau, of New Orleans, of a girl who at birth was as much developed as a girl of fourteen years. At the age of three years her period came on and continued naturally every month, lasting each month three days. Such cases are of course very unusual.

The period at which this change to puberty takes place is marked by a series of changes which show the revolution undergone by the system. For some time previous the nervous system has felt the change; the temper becomes variable, at times uncertain. A girl who before was probably gay and boisterous in her deportment becomes timid and shy, easily embarrassed, the slightest cause making her blush, the knowledge of which may add to her embarrassment. She may notice a gradually increasing development of her figure, which annoys her and makes her shun company. Her younger companions no longer have the same charm; involuntarily she prefers the society and dress of those older than herself. The watchful mother at once recognizes the cause which is bringing about these changes, and it is her duty to gain the confidence of her child and, without exciting her suspicions or disturbing her already uncertain nervous system, lead her to understand that she is no longer a child. She is a woman, and it falls to the lot of all women who are in good health to have a certain monthly drain upon their system which is calculated to relieve the other organs of the body; and should it appear at any time, which it is now likely to do, she should avoid all things that might check it, take certain precautions, as rest, etc., to secure its regular return, which should be painless. As this often in many cases comes unawares, certain precautions in clothing have to be taken; especially is this the case if the child should happen to be away from home at this time. Then again it is well for them to learn that the menstrual flow may at times, especially in delicate girls, be attended by severe pain, by nervous prostration, in fact by a constitutional thunder-storm. There may be severe headache (frequent in such cases), colic,—this latter may be extremely severe; there may be giddiness, nausea, extreme nervousness, chills or creeps, excessive backache, all of which may come on suddenly, following some slight indiscretion in diet which tends to mislead the person from its true cause.

When a young girl once distinctly understands that the object of her monthly flow is to keep her in good health,—and this surely is the light in which it should be presented to her,—she will readily understand that to secure the health which is heaven's greatest blessing she must be strictly guarded as to the care of herself. She should understand that after one appearance the periodicity of this function is not at once established; there is frequently great irregularity in its return. It may last but a few hours and then return in fifteen days; it may be copious at first and then not return for two months or more. Instead of appearing as it should at the time of its expectation, it may appear as a bleeding from the nose, it may be replaced by a diarrhœa, or by a discharge devoid of color.

There are certain signs which note the advance of a period, and it is well for the mother to impress upon her daughter to be careful of them. Although the normal return of the period is calculated by the lunar month, really the idea that the moon exerts an influence upon this condition has no basis; this is shown by the fact that the day of return differs in women. Again, some have the normal return every thirty, others every twenty-five or twenty-eight days. The duration of the period also differs in most women, some lasting three or four days, others eight or more. The quantity also differs; the range comprised within the area of health is widely spread. These facts are important to bear in mind; though should any deviation exist from that which the mother believes to be the normal condition, she should mention the fact to the family physician, and let him be the judge of its importance.

During this period of life there is a tendency in certain types of young girls to develop what is known as hysterical phenomena. Now, it is well to know that what is understood as hysteria by physicians is not merely those attacks which people call hysterics.

Physicians frequently hear this reply, "But, doctor, my daughter is not hysterical; she is of the most amiable disposition; I know her to be extremely brave and fearless." At the same time she may be a marked subject of hysterical vomiting, or some of the paralyses. It is a difficult matter to decide whether these troubles—which are usually termed reflex by doctors, because they are sympathetic or reflected from other organs through the chain they have in common, the delicate nervous system—are due to disturbance in the womb, or are simply the result of an associated weakness, of which the disturbed menstruation, the pain, or diminished quantity of the flow is another evidence.

In many cases the womb-troubles, which may simply be the irregularity of the function of menstruation, may be the cause, and rest, hot foot-baths, electricity, etc., bring about a cure; or it may be due to weakness on the part of the individual, poor blood, deficient outdoor employment, too much standing, as is so common with store-girls, and only yield to tonics, fresh air, ample diet, and exercise.

What are generally understood as hysteria by the non-professional are the outbursts from the nervous system upon the slightest irritation, whether pleasurable or painful. To a certain extent this is independent of any disease or even disorder of the generative system, and is solely, I regret to have to say, due to bad "bringing-up." Gentle, over-indulgent parents are themselves the cause of such a state of affairs.

It is not for us to study the cause of these changes in individuals that produce this function, nor to describe the anatomy and physiology of the organs that are engaged in it. It is merely necessary to insist upon the fact that normal menstruation should be painless, and that disturbances occurring at this time, whether in the form of local pain, headache, or lassitude, bear the same relation to normal menstruation that discomfort, nausea, and pain bear to healthy digestion. Just as dyspepsia is dependent upon indiscretion in diet, or weakened digestion from debility, so difficult or painful menstruation is the result of indiscretion at the time when care and watchfulness should be the rule, or it is the penalty paid for neglect at some earlier period.

Debility in early girlhood is one of the principal causes of pain when the function is established. It is usually found in rapidly-growing girls whose tastes have led them enthusiastically into literary pursuits, partly from a feeling on their part that their muscular weakness will prevent their taking pleasure in the rough out-door pleasures of their more robust companions, and partly from the extreme excitability of their nervous system, which makes them at an early age noted for their brilliancy, and which also will exaggerate their liability to pain. Such children are easily recognized, and to them the watchful mother should give her careful attention in anticipation of what their development will bring forth. Their studies should be gently directed towards those pursuits that lead them out of doors; the muscular exercise involved in household duty should be gradually given them; habits of early rising and early to bed should be insisted upon. There should be a judicious division of their school-hours, the selection of well-ventilated and bright school-rooms, daily gymnastic exercise, swimming, riding on horseback, and walks,—not the aimless promenading the streets, but walks that are calculated to give both pleasure and profit. There is no better way of making healthy girls than to make the various branches of science according to their taste a part of their education; there is not a girl, or in fact any individual, who has not a latent taste which cannot be with a little care developed. If fond of drawing or painting, encourage it from the earliest moment; teach her to draw from nature, and she will spend hours in the open air. Botany, mineralogy, photography, will embrace the repertoire of an educated woman as well as Latin and Greek, and a woman's mind is capable of accommodating them all, if necessary.

Unfortunately, the present fashion is totally at variance, strange to say, with that of the ancient days of Greece and Rome. Now a person can learn nothing except in an ill-ventilated school-room, in a barrel-hoop position. The philosophers of ancient days studied as hard as those of modern times, but they sought the solitude of the woods, and made their studies a pleasure instead of a task. It is not intense study that breaks so many down; it is the confined air, the sitting in a bent posture, the loss of appetite, the muscular weakness, and the poor blood that does the harm. I wish I could impress firmly upon mothers the importance of an erect carriage in young, growing girls. It is not merely the case that stooping shoulders and curved spines make their daughters unattractive in appearance, but that such conditions are absolutely a predisposing cause of disease; and I am satisfied from personal observation that such girls are always more or less affected with painful periods. To correct this before the age of puberty should be the aim of every mother. Make your girls from habit straight, with shoulders well thrown back, and they will avoid many womb-troubles in the future. If out-door exercise does not seem to correct this habit, there are certain movements of the muscles—light gymnastics—which your family physician can tell you of that are beneficial. One of the straightest and best-formed girls I ever saw, a picture of health, owed it all to her mother, who, noticing a gradual habit of stooping, required her to lie flat on her back on the floor without a pillow for one hour each day, while she read some entertaining book to her. Another matter which is important in this connection is the question of young girls wearing corsets. Fortunately, so much attention has been paid lately to the subject of dress in England, that those whose opinion is most valuable have freely expressed themselves. I cannot but fully endorse the statement of the London *Lancet* recently, that corsets should not be worn by young women; their dress should always be made so as to give free and independent movement to every part of the body; their garments should be light in weight, and the burden should fall as much as possible on the shoulders; there should be no restriction of any part of the body, and if a girl's figure needs a corset to make it shapely, let her endeavor to accomplish the same end by the more natural means of muscular development, which will at the same time give health as well as beauty. A strong back is far better than a corset.

Young girls should know that when the time comes

for their period they should avoid everything that would either postpone it or make it painful. If a horseback excursion has been fixed for that time, some excuse must be made; if a boating-party involve exposure to the night air, by all means avoid it. In avoiding extremes, one does not merely mean the extreme of doing too much, but also that of doing too little. The lounging about one's bedroom, or spending the whole day sitting reading a novel, will be as apt to give trouble as the opposite extreme; it will make the circulation sluggish, tend to headache, make the liver torpid, give rise to indigestion, and weaken the system. If one is accustomed to moderate exercise, the daily avocations of life should not be interfered with; but heavy lifting or over-fatigue in walking, or too long standing, riding on horseback, or dancing, should be carefully avoided.

I cannot lay too much stress on the importance of a careful regulation of the bowels, especially in its bearing upon the disorders of women. Every one knows the necessity of establishing a daily habit, from the fact that the wastes of the body which are discharged through the intestine, if allowed to remain, decompose, are reabsorbed and produce a certain poisoned condition of the system, made evident by furred tongue, nausea, distaste for food, disagreeable taste in the mouth, headaches, bad complexion, pimples, and other disorders of the skin, rendering the individual miserable to herself and others. Constipation also allows an accumulation to take place in the bowel, which by distending it will press upon the organs that lie in contact, and cause the extreme pain so common at times, especially in the left side. Nothing should interfere with this daily duty; but no one should use powerful purgatives without consulting a physician.

The varieties of disordered menstruation are known as *amenorrhœa*, *dysmenorrhœa*, and *menorrhagia*; the former meaning absence of menstruation, the second difficult or painful menstruation, and the third an unusual flow at that time. Delayed menstruation—that is, where a young girl has reached say the age of nineteen without the appearance of the flow—is apt to give rise to much anxiety. If this is associated with evidences of tardy sexual development, it is not of itself alarming, and cases are on record where women have married and given birth to children without ever having menstruated. But it is always well, and especially if occasional signs are present of an attempt at the establishment of menstrual flow, not to allow this state of affairs to run on long. The family physician should be consulted, as the obstacle may be a mechanical one, or certain causes exist which could be readily removed, but which otherwise might lead to serious disease. It is a recognized fact that the general mortality of women is increased at this period of life, and as soon as the establishment of the menstrual flow takes place the mortality shows a reduction. In the cases just mentioned it will be readily understood that to attempt to bring on the period by hot baths, poultices, or especially by the severe and forcible means of powerful drugs obtained without the doctor's consent, would be harmful, even fatal, and I regret to say that many such accidents happen. Dr. Mathews Duncan thus writes of delayed menstruation: "Like other processes of development, that of the generative system admits of considerable variation in point of time, without of necessity passing the limits of health. Indeed, just as one child cuts its first tooth at seven months and another not till a year old, so one girl will menstruate at fourteen to fifteen, and another not till seventeen." Weakness or feebleness of constitution, more or less the result of city life, may be in itself another cause for this condition. I quote again from the above author:

"A girl previously in good health approaches the

time of puberty, some of the changes characteristic of it take place: the form assumes the contour of womanhood, and nothing but the occurrence of menstruation is wanting to announce the completion of the change. The menses, however, do not show themselves, but the girl begins to suffer from frequent headaches; and the flushed face, frequent backache, pains in the lower portion of the abdomen, constipation, a furred tongue and a full pulse, and all these signs of constitutional disorder undergo a marked increase at stated periods of about a month. At length menstruation occurs, though in all probability scantily, and attended with much pain, and then for several months together there is no sign of its return. The general health was at first probably not seriously disturbed, but by degrees the patient becomes habitually ailing, the appetite falls off, the powers of digestion are weakened, the strength becomes unequal to ordinary exertion, the pulse grows feeble and frequent, and the face itself assumes a pallid, sallow tinge, whence the term 'chlorosis,' 'green sickness,' has been selected;" and it might be added that such patients are not by any means of necessity thinner than usual. The great mistake most people make is to attribute this to disorder of the liver; they call it biliousness, and are apt to do harm by overdosing for this supposed condition. Again, they will imagine languor represents weakness, and immediately have recourse to some strong preparation of iron or the inevitable dose of quinine, and then seem surprised that no improvement follows. It is the tonic influence of fresh air, healthful pursuits, exercise short of fatigue, and a nourishing, wholesome diet, of which milk should form the essential feature, that does most good, aided by those drugs which the investigation of the careful physician has warranted his suggesting to aid nature.

In the treatment of this form of difficult menstruation, which is applicable as well to those cases where from one cause or another the period has been missed by an ordinarily healthy girl, either from exposure to cold or the result of some shock to the system, it might be the debility consequent upon convalescence from fever, or some such cause, Dr. Duncan says, "The patient should be kept quiet, and if there is any considerable suffering or much disturbance of the circulation, it is desirable that she remain in bed, while the hot hip-bath night and morning, rendered still more stimulating in cases where the local pain is not considerable by the addition of some mustard." A gentle laxative should be administered, such as a dessertspoonful of the compound liquorice powder, or a dose of magnesia or phosphate of soda, or, better than all, probably, a dose of castor oil, preceded by a couple of Lady Webster pills at night. Should there be much pain, hot applications to the abdomen, either in form of a light meal poultice or a flannel bag of hops wrung out frequently in hot water, in addition to a hot foot-bath. Under no circumstances whatever should the powerful irritants sold in the drug-shops for such purposes be used; all drugs should be left for the physician to order. The use of hot teas is recommended,—ginger tea, tansy tea, etc.,—and by such means endeavor to encourage and not force the habit of menstruation. An excellent tea can be made from powdered ginger, senna, and dulcamara; it is laxative, and can be used every night with no bad effects.

The subject of pain during this time next claims our attention. At times it is so intense as to demand immediate relief; and frequently persons, especially those who have not the mother's care or her experience to guide them, will put off month after month the consultation with their physician, hoping that time will bring relief, and endeavor by various means to struggle through the times which to them bring renewed horrors, and finally wear them out with the constant effort to withstand the pain. And what means do they adopt to obtain relief? I regret to say it, little reflection is given to the cause of the disturbance. The period once over, they assume the same habits which have resulted in making a function normally painless fraught with pain; the same giddy life of society or the same over-indulgence in mental excitement and all that tends to enervate both body and mind. The cause of their trouble never for a moment attracts their serious attention; the treatment that they apply for their relief consists only in that which dulls their sensibility and deadens their nervous system. Alcohol in some form, whether gin or brandy, and opium, bromides, or chloral, are used in large amounts, and made to play the part of the greatest curse of a household.

I do not mean to say that all those who suffer do so on account of indiscretion, nor do I believe that the most rigorous and careful living would at once relieve the tendencies to periodical pain; but I cannot dwell forcibly enough upon the fact that those whose temperament is such that the slightest cause will result in hours of torture, can be in time relieved by rest, nutritious diet, and careful living.

If there be one cause more frequent than others to which the agony of the period is due, it is constipation; and I feel satisfied that nine out of ten sufferers owe their trouble to this cause. When the question is asked, they will positively assert that their bowels are moved daily with regularity; but probably a very small portion of the matter contained is passed, and a large amount remains accumulated, which presses against the tender, congested ovary as a morsel of food or a filling presses on the sensitive nerve of an inflamed tooth, and causes the severest form of neuralgia. I have not gone deeply into this subject, but have been sufficiently explicit to permit any one to understand this matter so as to prevent it. By the frightful abuse of stimulants, though the habit is brought about without a thought on the part of the sufferer, she is doing herself a most grievous wrong; month after month, each time increasing the dose, she will have recourse to her bottle of gin, her mixture of morphia, or her bottle of chloral, until finally she recognizes the unfortunate fact that she has become a victim to its use. It is far better to begin by a firm determination of avoiding them at the onset.

You want to relieve a congestion and bring on the flow. To do this, hot cloths or poultices to the abdomen, hot salt-bags to the back, hot hip-bath and foot-bath, rest in bed or lounging around one's room in loose clothing; or if the pain is not too severe, a moderate walk, or an agreeable change that will take the mind off themselves, and calm the nervous system by pleasant thoughts. The next thing should be the administration of a laxative, and probably the best of all is an enema which will secure the thorough removal of matter that may be the cause of trouble. A thin gruel may be used, made of oatmeal and strained. Of this about a quart should be used, or the ordinary Castile soap and water. If mothers who are solicitous about the well-being of their daughters would gain their confidence at these times, and minister to them during their hours of pain, applying those remedies which their experience has proved valuable, there would be far less suffering and far less danger of the habitual use of drugs that can be hidden in some convenient closet and taken *ad libitum*. Frequently the pain is so severe that a physician is called in, who gives a prescription, a strong anodyne in a pleasing mixture, or suppositories of opium. The relief is magic; the patient falls into a quiet sleep, and, barring the nausea and headache of the following day, is surprised at the result. A copy of the prescription is obtained, and it serves ever after for herself and friends. This is wrong, and every right-minded woman should feel that to expose herself to a habit of this kind is to sacrifice her life to a slavery which ends only in the grave. It is far better to consult the family physician at once, tell him frankly and without hesitation what the trouble is. It is an old story to him; he has listened to many lectures upon it, he has recited your symptoms in class, he has heard the tale often told, until it has assumed a very monotonous sound. You imagine that the matter is too delicate a one to speak of without embarrassment, but you forget that the position your family physician holds towards you is one so intimate and confidential, so sacred in its association, that it has received the sanction of heaven itself, as you are bidden to obey him.

The severest form of painful menstruation is that which occurs in young women whose period has not appeared till much later age than usual. "The pain in such cases precedes menstruation for a day or two, generally reaches its greatest intensity in the course of the first thirty-six hours of the flow, being sometimes so intense that the patient writhes in agony, and then often by degrees subsides, though it does not cease entirely till the period is over, though severest in the uterine and pelvic regions (lower part of the abdomen). The pain is not generally limited to these situations, but is experienced also in the back and loins, and shoots down the inside of the thigh. The pain, too, is aggravated at intervals, and becomes paroxysmal like colic; and the whole abdominal surface is so tender as scarcely to bear the slightest touch. Intense headache is very frequent, often confined to one side of the head; and in other cases the stomach is disordered and the patient distressed by constant nausea and vomiting." The treatment may be summed up as follows: absolute rest before the period is expected, avoidance of any active enterprise, running up and down stairs, horseback exercise, tennis, long walks, and sudden changes of temperature; to have the bowels moved freely by compound liquorice powder or some such simple laxative. Exposure to cold is very apt to bring this on, and it is especially liable to occur in girls who sit out of doors after dark with their thin summer clothing, and allow the damp, cool air after nightfall to chill the surface before the expected period. This is frequently noticed at the sea shore. How many of the gayly-dressed beauties, in their lightest clothing, will dance a waltz through and then rush frantically for a walk on the porch, and will next day suffer tortures in their rooms for their imprudence, while their friends marvel at the number of sick headaches they seem to have. Sleep, rest,—absolute rest in bed,—hot foot-baths prolonged and frequently repeated, a strong, hot, well-seasoned cup of beef-tea, is the best means of procuring relief. If the pain still continue, a hot lemonade with a teaspoonful of sweet spirits of nitre to the tumbler. The applications to the surface of the abdomen are usually valuable on account of their warmth; a mush poultice well sprinkled with laudanum is useful, or a bag of hops quilted and wrung out frequently in hot water and saturated with laudanum, or wet with the spirits of chloroform. Frequently a hot salt-bag to the back will give relief, or the rubber bag filled with hot water. Should the pain still remain severe notwithstanding these domestic remedies, the physician should be sent for and the matter fully explained to him.

There is another form of painful menstruation which is less apt to occur at an early period of life, and is recognized by the following symptoms, according to Dr. Duncan: "During the first twenty-four or thirty-six hours of each menstruation the discharge in general is but scanty, and the pain is very severe. At the end of this time, however, sometimes sooner, the hemorrhage often becomes abundant, and as the blood flows the pain again abates and then ceases altogether. In some of the cases, the discharge having continued for a few hours ceases, and then comes on again, while though scanty it is intermixed with small 'clots.'"

In these cases anodynes no longer furnish the ready relief which follows their administration in the neuralgic form. There are so many causes for this form of dysmenorrhœa, that the physician should at once be consulted, in order that the proper treatment may be instituted before the next period. Laxatives, such as the various purgative waters, give relief in these cases; the granular effervescent Carlsbad salt, a tablespoonful in water, may be taken, or a claret glass of Hunyadi water, or Friedrichshall water with an equal part of hot water. A free purgative action of the bowels should be the first thing to accomplish. If a patient is away from a physician, she should take a tablespoonful of liquor ammoniæ acetatis during the painful time, when the flow is scanty, in some weak hot lemonade every two hours, until three or four doses have been taken.

It is so obviously the mother's duty to be the confidant of her child while nature is establishing these physiological processes, that I have felt the necessity of adding this chapter to my book. Most of the disorders of the nervous system which attend this time of life are directly traceable to indiscretions which are the result of ignorance. The intellectual girl of sedentary habit offends nature by retarding her physical development. The thoughtless enthusiast lays the seed of future disorders by an ignorance which the timely admonition of a watchful mother would avoid.

INDEX.

Air, 87, 91.
Air-passages, diseases of, 151.
Amenorrhœa, 210.
Anodynes, abuse of, 95, 214.
Aperients, 45, 124.
Aphthæ, 111.
Atomizer, 133, 153.
Attitudes, unhygienic, 209.

Baby-basket, 34.
Barley as infants' food, 71.
Bathing, 94, 97.
    during pregnancy, 21.
Binder, 35, 39.
Bone, 102, 195.
Bottle, 55, 66.
Bottle-feeding, 50, 66, 85.
Bowels, 116.
Breasts, care of, 32.
    changes in, during gestation, 14.

Calisthenics for children, 199.
    for girls, 203.
Carnrick's Soluble Food, 81.
Catarrh, nasal, 127.
    of stomach and bowels, 168.
Cerumen, 141.
Childhood, 126.
Child's nurse, 94.
Chlorosis, 202, 211.
Cholera infantum, 168, 169.
Cleanliness, 186.
Clothing, children's, 198.
    during pregnancy, 31.
    infantile, 40.
    of young children, 199.
Colds, 87, 127.
Colic, 124.
Condensed milk, 59.
Confinement, 14, 34.
    probable date of, 13.
Congestion, how to relieve, 215.

Constipation in pregnancy, 23.
    effect of, on nursing child, 44.
    infantile, 116.
    as a cause of uterine disease, 210.
    as a cause of monthly pain, 214.
Convulsions, infantile, 110.
Corsets for girls, 209.
    in pregnancy, 30.
Cows' milk, 52, 66.
Cramp, how to relieve, 26.
Cravings in pregnancy, 18, 21.
Creeping, 105.
Croup, 151, 155, 157.

Date of confinement, how to compute, 13.
Debranned flour, 70.
Dentition, 100, 112.
    second, 191, 201.
Diapers, 198.
Diarrhœa, 164.
Diet before confinement, 21.
    after confinement, 42.
    after early dentition, 112.
Digestion, 66.
Diphtheria, 151, 157.
Disinfectants, 159, 163, 185.
Douche, nasal, 132.
Dress in pregnancy, 30.
    of children, 198.
    of infants, 40.
Dysmenorrhœa, 210, 216.
Dyspepsia in pregnancy, 17.

Ear, diseases of, 135.
Earache, 135, 143.
Education, 208.
Emotions, their effect on offspring, 33.
Enemata, 25, 121.
    in pregnancy, 25.

INDEX.

Epistaxis, 133.
Exercise before confinement, 27.
　children's, 87.
　for young girls, 208
Eye, diseases of, 144.

Fairchild on peptonized food, 74.
Farinaceous foods, 82.
Flour-ball, 69, 71.
Food after dentition, 115, 197.
　infants', 50, 56, 65, 69, 80.
Foods, prepared commercial, 80.
Foot-bath, 112.

Gastric juice, 67.
Gestation, the troubles of, 14.
　diet in, 21.
　not a disease, 16.
Girls at first menstruation, 200.
Gum-arabic water as food, 82.
Gums, lancing, 111.

Hardening of children, 155.
Hygiene of gestation, 21.
Hysterical symptoms, 206.

Ideal mother, 34.
Impressions, their effect on offspring, 33.
Indigestion of pregnancy, 17.
　of infancy, 70.
　of childhood, 122.
Infancy, 38.
Infant, 40.

Lactation, 42.
Laxatives, 45, 124.
　at menstrual period, 212.
Leeds's, A. R., preparation of milk, 73.
Lemonade of peptonized milk, 76.
Liebig's foods, 82.
Lime-water, 57, 65.

Malt liquors in lactation, 43.
Mammary glands, 32.
Maternity, 13.
Measles, 186.
Meigs, A. V., his preparation of milk for the bottle, 64.
Membranous croup, 157.

Menorrhagia, 210.
Menstruation, delayed, 216.
　disordered, 210.
　first, 203, 211.
　painful, 213, 216.
Milk, boiled, 72.
　condensed, 59.
　cows', 52, 66.
　" evaporated," 59.
　humanized, 78.
　mother's, 42.
　mother's, as affected by diet and medicines, 43.
　peptonized, 75.
　to check the flow of, 43.
Milk-foods, 80, 81.
Milk-jelly, peptonized, 76.
Miscarriage, 36.
Monthly nurse, 35.
Morning sickness, 19.
Mother, the ideal, 34.

Nasal catarrh, 127.
　douche, 132.
Nausea in pregnancy, 19.
Navel, care of, 39.
Nestle's Food, 81.
New-born infant, 38, 40.
Nipple for nursing-bottle, 55.
Nose-bleed, 133.
Nose, blowing the, 131, 137.
Nurse, child's, 93.
　monthly, 35.
Nursery, 88.
Nurses, wet-, 45.
Nursing, 48.
Nursing-bottle, 55, 66.

Oatmeal, 71, 85.
Ophthalmia, 144.
Out-door exercise, 87.
Over-feeding, 108, 164.

Pain, how treated, 213.
Pancreatic extract, 74.
Parker, W. T.; his method of preparing milk, 58.
Peptogenic Milk Powder, 73.
Peptonization, 74.
　of milk, 75, 77.
Perambulators, 93.
Plants, 92.
Pond's Extract, 134.

INDEX.

Pre-digestion, 73.
Pregnancy, 13.
   diet of, 21.
   is not a disease, 16.
Pregnant woman's duties towards her offspring, 33.
Premature labor, 36.
Puberty, 200.
Purgatives during gestation, 2.
   in dysmenorrhœa, 216.

Recreations of young girls, 203.
Rickets, 102, 196.
Running from the ear, 138.

Salt-bath, 97, 105.
Scarlatina, 180.
Scarlet fever, 180.
Sea-shore, 97.
Sewerage pipes, 90.
Sleep of infants, 88.
Sleeping-room, 90.
Sleeplessness, 106, 109.
Smith's (J. L.) formula for infants' food, 69.
Snoring, 132.
Soap, 99.
Soothing medicines for infants, 95.
Soup, peptonized, 76.

Starchy foods for infants, 70.
Stimulants to be avoided, 214.
Syringe, traveller's, 25.

Tainted milk, 51.
Teaspoon, how much it holds, 57.
Teething, 100, 112.
Throat, diseases of, 151.
Toilet appliances for the lying-in room, 34.
Traveller's syringe, 25.
Twelfth-year molar teeth, 20.

Under-feeding of infants, 107.

Valsalva's experiment, 137.
Vaporizer, use of, 133, 153.
Ventilation, 87, 89.
Vichy water in infants' milk, 61.
Vomiting in indigestion, 123.
   in infantile diarrhœa, 173.
   in pregnancy, 19.

Water for infants, 165.
Wax in the ears, 141.
Weaning, 83.
Wet-nurses, 45.
Whey as food, 65, 66.
Whooping-cough, 174.

THE END.

A LIST OF BOOKS

SELECTED FROM THE

Catalogue

—OF—

J. B. LIPPINCOTT COMPANY.

(COMPLETE CATALOGUE SENT ON APPLICATION.)

Medical.

P. H. Maclaren, M.D., F.R.C.S.E.
Atlas of Venereal Diseases. Complete in 10 Parts. Containing 30 Colored Plates and over 40 Subjects. Paper, royal 4to, $2.00 per part.

Pliny Earle, A.M., M.D.
The Curability of Insanity. A Series of Studies. 12mo, cloth, $2.00.

E. Landolt, M.D.
The Refraction and Accommodation of the Eye and their Anomalies. Translated by C. M. Culver, M.A. With 147 Illustrations. 8vo, cloth, $7.50.

Philip Schech, M.D.
Diseases of the Mouth, Throat, and Nose. Including Rhinoscopy and the Methods of Local Treatment. Translated by R. H. Blaikie, M.D., F.R.C.S.E. With Illustrations. 8vo, cloth, $3.co.

Practical Lessons in Nursing.
Each 12mo, cloth, $1.00.

I.—The Nursing and Care of the Nervous and the Insane. By Chas. K. Mills, M.D.

II.—Maternity; Infancy; Childhood. The Hygiene of Pregnancy; the Nursing and Weaning of Infants; the Care of Children in Health and Disease. By John M. Keating, M.D.

III.—Outlines for the Management of Diet; or, The Regulation of Food to the Requirements of Health and the Treatment of Disease. By E. T. Bruen, M.D.

Prof. C. Schweigger.

Hand-Book of Ophthalmology. Translated by Dr. Porter Farley. Illustrated. 8vo, cloth, $4.50.

Thomas's New Medical Dictionary.

A Complete Pronouncing Medical Dictionary. Embracing the Terminology of Medicine and the Kindred Sciences, with their Signification, Etymology, and Pronunciation. With an Appendix, comprising an Explanation of the Latin Terms and Phrases occurring in Medicine, Anatomy, Pharmacy, etc., together with the Necessary Directions for Writing Latin Prescriptions, etc., etc. By Joseph Thomas, M.D., LL.D. Imperial 8vo, 844 pages, extra cloth, $5.00. Library sheep, $6.00.

A Comprehensive Pronouncing Medical Dictionary. Containing the Etymology and Signification of the Terms Made Use of in Medicine and the Kindred Sciences. Crown 8vo, cloth, $3.25 Sheep, $3.75.

Practice of Pharmacy.

With over 900 Pages and nearly 500 Illustrations. A treatise on the Modes of Making and Dispensing Officinal, Unofficinal, and Extemporaneous Preparations, with Descriptions of their Properties, Uses, and Doses. By Joseph P. Remington, Ph.G. Intended as a Hand-Book for Pharmacists and Physicians, and a Text-Book for Students. 8vo, cloth, $5.00. Sheep, $6.00.

Wormley's Micro-Chemistry of Poisons.

Including their Physiological, Pathological, and Legal Relations. With an Appendix on the Detection and Microscopic Discrimination of the Blood. Adapted to the Use of the Medical Juris., Physician, and General Chemist. By Theodore G. Wormley, M.D., Ph.D., LL.D. *A New, Revised, and Enlarged Edition.* With new Illustrations. 8vo, cloth, $7.50. Sheep, $8.50.

J. E. Garretson, M.D., D.D.S.

A System of Oral Surgery. Being a Consideration of the Diseases and Surgery of the Mouth, Jaws, and Associate Parts. *Fourth Revised and Enlarged Edition.* Illustrated. 8vo, cloth, $8.00. Sheep, $9.00. Half Russia, $9.50.

F. H. Gerrish.

Prescription Writing. For the Use of Students who have never studied Latin. 18mo, cloth, 50 cents.

Wm. A. Hammond, M.D.

Lectures on Venereal Diseases. 8vo, cloth, $2.00.
Sleep and its Derangements. 12mo, cloth, $1.75.
Physiological Essays. 8vo, cloth, $2.00.

Hand-Book of Nursing.

For Family and General Use. 12mo, cloth, $1.25.

Dr. Carl Hoppe.

Percussion and Auscultation as Diagnostic Aids. Translated by L. C. Lane, M.D. 12mo, cloth, 50 cents.

Thomas Kirkbride, M.D.

Hospitals for the Insane. *Second Edition, Enlarged and Revised.* Illustrated. 8vo, cloth, $3.00.

O. F. Manson, M.D.

A Treatise on the Physiological and Therapeutic Action of the Sulphate of Quinine. 12mo, cloth flexible, $1.00.

Laurence Turnbull, M.D., Ph.G.

Clinical Manual of Diseases of the Ear. *Second Revised Edition.* With 114 Illustrations. 8vo, cloth, $4.00.
Imperfect Hearing and the Hygiene of the Ear. *Third Edition.* With Illustrations. Large 8vo, cloth, $2.50.

Lewis A. Sayre, M.D.
Spinal Diseases and Spinal Curvature. Illustrated with Photographs from Nature. 12mo, cloth, $4.00.

A Dictionary of Practical Surgery.
By Various British Hospital Surgeons. Edited by Christopher Heath, F.R.C.S. 2000 pages. 8vo, cloth, $7.50. Sheep, $8.50.

William Stirling, M.D.
A Text-Book of Practical Histology. With 30 Outline Plates, 1 Colored Plate, and 27 Wood Engravings. Quarto, cloth, $4.50.

Drs. T. G. Morton and Wm. Hunt.
Surgery in the Pennsylvania Hospital. With Papers by J. B. Roberts, M.D., and F. Woodbury, M.D. Illustrated. 8vo, cloth, $4.00.

P. H. Chavasse, M.D.
Advice to a Wife, and Advice and Counsel to a Mother. Advice to a Wife on the Management of her own Health, and on the Treatment of some of the Complaints incident to Pregnancy, Labor, and Suckling. Advice to a Mother on the Management of her Children, and on the Treatment of some of their more Pressing Illnesses and Accidents, etc. Three volumes in one. *New Edition.* 12mo, cloth, $2.00.

United States Dispensatory.
Edited by Prof. H. C. Wood, M.D., Prof. Joseph P. Remington, Ph.G., and Prof. Samuel P. Sadtler. *Fifteenth Edition, Thoroughly Revised, Enlarged, and Corrected.* Illustrated. 8vo, cloth, $7.00. Sheep, $8.00. Half Russia, $9.00. Any of the above styles, with Denison's Patent Index, additional, 75 cents.

Worcester's Dictionary
THE STANDARD
IN SPELLING, PRONUNCIATION, AND DEFINITION.

New Edition. The largest and most complete Quarto Dictionary of the English Language. 2126 pages. Contains thousands of words not to be found in any other Quarto Dictionary. Enlarged by the addition of A BIOGRAPHICAL DICTIONARY of nearly 12,000 personages, and A GAZETTEER OF THE WORLD, noting and *locating* over 20,000 places. Containing also over 12,500 new words, recently added, together with a table of 5000 words in general use, with synonymes. Illustrated with wood-cuts and full-page plates. With or without Denison's Patent Index.

SCHOOL DICTIONARIES.

WORCESTER'S PRIMARY DICTIONARY.
Profusely Illustrated. 384 pages. 16mo. Half roan

WORCESTER'S NEW SCHOOL DICTIONARY.
With numerous Illustrations. 390 pages. Half roan.

WORCESTER'S COMPREHENSIVE DICTIONARY.
Profusely Illustrated. 688 pages. 12mo. Half roan.

WORCESTER'S NEW ACADEMIC DICTIONARY.
688 pages. 12mo. Half roan.

WORCESTER'S are *the latest school dictionaries published;* they give the correct usage in *pronunciation;* they contain *a much larger number of words* than any other school dictionary ; they give the correct usage in *spelling;* the *definitions* are complete, concise, and accurate.

FOR SALE BY ALL BOOKSELLERS.

J. B. LIPPINCOTT COMPANY,
PUBLISHERS,
715-717 MARKET STREET, PHILADELPHIA, PA.

www.ingramcontent.com/pod-product-compliance
Lightning Source LLC
Chambersburg PA
CBHW021843230426
43669CB00008B/1062